Innovative brain tumor therapy

Related titles

NMR metabolomics in cancer research
(978 1 907568 84 8)

Nanoparticulate drug delivery: Perspectives on the transition from laboratory to market
(978 1 907568 98 5)

Electroporation-Based Therapies for Cancer: From basics to clinical applications
(978 1 907568 15 2)

Woodhead Publishing Series in Biomedicine: Number 67

Innovative brain tumor therapy

Nanoparticle-based therapeutic strategies

GERARDO CARUSO, LUCIA MERLO, AND MARIA CAFFO

ELSEVIER

AMSTERDAM • BOSTON • CAMBRIDGE • HEIDELBERG • LONDON
NEW YORK • OXFORD • PARIS • SAN DIEGO
SAN FRANCISCO • SINGAPORE • SYDNEY • TOKYO
Woodhead Publishing is an imprint of Elsevier

WP WOODHEAD PUBLISHING

Woodhead Publishing is an imprint of Elsevier
80 High Street, Sawston, Cambridge, CB22 3HJ, UK
225 Wyman Street, Waltham, MA 02451, USA
Langford Lane, Kidlington, OX5 1GB, UK

British Library Cataloguing-in-Publication Data
A catalogue record for this book is available from the British Library.

Library of Congress Control Number: 2014946707

ISBN 978-1-907568-59-6 (print)
ISBN 978-1-908818-74-4 (online)

For information on all Woodhead Publishing publications visit our website at http://store.elsevier.com/

Typeset by RefineCatch Limited, Bungay, Suffolk

Transferred to Digital Printing in 2014

Contents

List of figures

List of abbreviations

AMT	adsorptive-mediated transcytosis
AON	antisense oligonucleotides
AQPs	aquaporins
BBB	blood–brain barrier
BCNU	1,3-bis(2-chloroethyl)-1-nitrosourea
CED	convection-enhanced drug delivery
CMT	carrier-mediated transport
CNs	carbon nanotubes
CNS	central nervous system
COX-2	cyclo-oxygenase-2
DARC	Duffy antigen receptor for chemokines
ECM	extracellular matrix
EGFR	epidermal growth factor receptor
EPR	"enhanced permeability and retention effect"
ERK	extracellular signal-regulated kinase
FGFR	fibroblast growth factor receptor
GBM	glioblastoma
GNPs	gold nanoparticles
GO	graphene oxide
GPEI	polyethyleneimine-modified magnetic NPs
GRP	gastrin-releasing peptide
GSPID	graphene nanosheet-based system
HB-EGF	heparin-binding epidermal growth factor
HER2	receptor tyrosine-protein kinase erbB-2
HIF-1	hypoxic ischemic factor-1α
HSP 90	heat-shock protein 90

IL-1β	interleukin-1β
IL-13R	interleukin-13 receptor
IP	inositol monophosphate
LDL	low-density lipoprotein
LNCs	lipid nanocapsules
MEK	mitogen-activated protein kinase kinase
MMPs	matrix metalloproteases
MNPs	micelles nanoparticles
MRI	magnetic resonance imaging
MRP	multidrug resistant protein
MSCs	mesenchymal stem cells
MWNT	multi-walled nanotubes
NGO	nano-graphene oxide
NIR	near-infrared
NPs	nanoparticles
NSCs	neural stem cells
PAT	photoacoustic tomography
PCL	poly(ε-caprolactone)
PDGFR	platelet-derived growth factor receptor
PDT	photodynamic therapy
PEG	polyethylene glycol
PLGA	polyglycolide
QDs	quantum dots
RMP-7	lobadimil
RMT	receptor-mediated transport/transcytosis
ROS	reactive oxygen species
siRNA	small interfering RNA
SPIO	superparamagnetic iron oxide
SWNT	single-walled nanotubes
TGF-β_2	transforming growth factor-β_2
TiO_2	titanium dioxide
TMZ	temozolomide
VEGFR	vascular endothelial growth factor receptor
ZO	zonula occludens

About the authors

Dr Gerardo Caruso (University of Messina, School Medicine, 1990) is currently an Assistant Neurosurgeon, Neurosurgical Clinic, University of Messina, Italy. He has been a Member of the Italian Society of Neurosurgery since 1996 and a Member of the American Nanosociety since 2011. Dr Caruso previously served as an officer in the Italian Navy and has also served as a reviewer for a number of international technical journals. He is author of various papers refereed on international scientific journals.

Dr Lucia Merlo (University of Messina, School of Medicine, 2001) is Neurosurgeon at the Neurosurgical Clinic, University of Messina, Italy. She has been a Member of the European Association of Neurosurgical Societies since 2010. She is interested in Neuro-oncology, Pediatrics, and Functional Neurosurgery.

Dr Maria Caffo (University of Messina, School Medicine, 1994) is currently an Assistant Neurosurgeon, Neurosurgical Clinic, University of Messina, Italy. She has been a Member of the Italian Society of Neurosurgery since 2000, a Member of the Italian Association of Neuro-oncology since 2000, and a Member of the European Association of Neuro-oncology since 2002. Dr Caffo has a PhD in Neuro-oncology. She is also Assistant Professor, Department of Neuroscience, Neurosurgical Clinic, University of Messina, Italy, and a reviewer for several technical journals.

Dr Caruso can be contacted at:
gcaruso@unime.it

Dr Merlo can be contacted at:
l.merlo@hotmail.com

Dr Caffo can be contacted at:
mcaffo@unime.it

Introduction

Despite recent advances in surgery as well as radiotherapy and chemotherapy, therapeutic efforts have not successfully established a definitive strategy of effective treatments for brain gliomas. Overall mortality is still high with an average survival less than 1 year in cases of glioblastoma.

The standard protocol includes maximal surgical resection with postoperative combination of radiotherapy with concomitant and adjuvant chemotherapy. This treatment has improved the median overall survival from 6 to 14.6 months. Notwithstanding, malignant brain tumors remain lethal.

The presence of the blood–brain barrier selectively impedes the passage of numerous types of molecules inside the brain. This limits the efficacy of current drugs. The understanding of epigenetic modifications during neural differentiation and an insight into oncogenesis processes will likely help in developing new epigenetic/genetic-based treatments acting upstream.

These complex fields may reveal the most effective therapeutic modalities but need selective drug delivery systems to reach the tumor cells. Hence, the new frontier in reliable brain tumor therapy places it trust in nanoparticles. Their nanometric size, electrostatic charge, and lipophilic characteristics allow them to penetrate into the brain tissue freely. Nanotechnology could therefore be used both to improve treatment efficacy and to reduce the adverse side effects. Nanotechnology-based approaches to targeted delivery of drugs across the blood–brain barrier may potentially be engineered to carry out specific functions as needed. Moreover, nanoparticles show tumor-specific targeting and long blood circulation times, with consequent low short-term toxicity. Promising in vitro results have been reported, but remain to be validated in humans. Nanomedicine, the application of nanotechnology to healthcare, holds great promise not only for revolutionizing medical treatments, but also in imaging for a faster diagnosis, and in tissue regeneration.

This book focuses on the pathophysiology of the blood–brain barrier in brain tumors, and the possibilities of overcoming this with nanoparticle-based systems. It presents a synopsis of the latest studies on nanoparticles as ideal devices for brain tumor treatment. Relevant patents of nanoparticles used as drug delivery carriers are also reported, as well as future scenarios concerning nanoparticles and stem cells.

Brief introduction on brain tumor epidemiology and state of the art in therapeutics

DOI: 10.1533/9781908818744.1

Abstract: This chapter focuses on the epidemiology, epigenetics and therapy of brain tumors. Standard treatment includes maximal surgical resection with postoperative radiotherapy and concomitant and adjuvant chemotherapy. Even with the most advanced multimodal standard treatments malignant brain tumors, namely gliobastomas, still have a poor prognosis. Numerous studies have been performed, and many are still underway, to develop successful treatments, and a major challenge seems to be defining the real cellular origin of this tumor. Starting from the initial cellular mutations and epigenetic modifications it will be possible to find new epigenetic-based therapies. We also stress the importance of overcoming the blood–brain barrier in order to improve the therapeutic effects.

Key words: brain tumor, epidemiology, epigenetics, glioblastoma, treatment.

The median worldwide incidence of primary central nervous system tumors has been estimated as 3.9 per 100 000

Published by Elsevier Limited, 2014

person-years,[1] gliomas accounting for 77% of the malignant subtypes.[2] The Central Brain Tumor Registry of the United States annual report on the 10-year period 1985–1994 revealed a slight (0.9%) but statistically significant annual increase in incidence,[3] indicating how this disease is gradually increasing in the developed countries. In fact, there is around a fourfold difference between countries with a high incidence (e.g. Australia, Canada, Denmark, Finland, New Zealand, and the USA) and regions with low incidence (e.g. the Philippines and India).[4,5] Differences in diagnostic practices, medical care, and completeness of reporting make these geographic comparisons difficult.

The term "glioma" encompasses all the neoplasms originating from mutated neural stem cells, and classified by the World Health Organization into four grades of malignancy,[6] starting from tumors with low proliferative potential and the possibility of being cured (e.g. pilocytic astrocytoma – grade I), to diffusely infiltrative astrocytic tumors with cytological atypia (e.g. diffuse astrocytoma – grade II, which are able to progress to higher grades of malignancy), to lesions with histological evidence of malignancy including nuclear atypia, anaplasia, and brisk mitotic activity (e.g. anaplastic astrocytoma – grade III), and finally to the cytologically malignant, mitotically active, necrosis-prone neoplasms, rich in microvascular proliferation, which are glioblastoma (GBM) – the highest grade tumor (grade IV).[6]

Patients with GBM have the highest median age at diagnosis and the worst prognosis. Survival time for people with malignant brain tumor is related not only to the histologic type of the tumor but also to age at diagnosis,[7] Karnofsky Performance Score status,[8] extent of surgical resection,[9] ki-67 immunohistochemical markers,[10] and sensitivity to chemotherapy as determined by genetic mutations such as *IDH1*,[11] *PTEN*,[12] *EGFR* amplification,[13] and *1p19q* codeletion.[14] Estimates of median overall survival vary widely and range from 4.4 to 65.5 months[15] for anaplastic astrocytomas whereas only 2% of patients aged 65 years or older, and only 30% of those under the age of 45 years at GBM diagnosis, survive for 2 years.[3]

To develop a successful treatment for GBM, a major challenge is defining the real cellular origin of this tumor. Although cells acquiring a mutation, like the tumor cells, may not be the same as the cell of origin,[16] epigenetic modifications, enzymes, and noncoding RNAs are often cell-type specific and can aid in identification of the cell of origin so as to focus therapies directly on it. In a mouse glioma model proposed by Liu et al.[17] the neural stem cells pass on mutations to downstream progeny such as oligodendrocyte precursor cells (OPCs), which become putative glioma cells of origin. By contrast, Koso et al.[18] suggested that the cell of origin in some GBM is not an OPC but an astroglial-like cell, and that the originating mutations can occur in neural stem cells. Hence, it may be likely that multiple cells of origin give rise to gliomas. Further, studying the genetic and epigenetic landscape of human OPCs and other cells as they are differentiating could uncover epigenetic enzymes and pathways misregulated in gliomagenesis. As an example, recent studies suggest that epigenetic modification determination during neural differentiation will probably provide insight into dedifferentiation processes, which may give rise to GBM. Friedmann-Morvinski et al.[19] demonstrated that mature neurons dedifferentiate and become tumor-initiating cells in mouse models of glioma. Parallel studies showed that targeting astrocytes promotes glioma formation.[20] Collectively, these findings suggest that gliomas may arise from either dedifferentiating neural stem cells or astrocytes. Importantly from a therapeutic point of view, dedifferentiating neurons or astrocytes that give rise to gliomas might be the very cells that are resistant to chemotherapy with temozolomide (TMZ) and that induce tumor recurrence.[21]

Moreover, recurrent genomic regions of alteration, net gains and losses, and DNA aberrations have been found as markers of gliomagenesis inside tumor samples. Whereas some of these regions contain known oncogenes and tumor suppressor genes, putative biologically relevant genes within other regions remain to be identified. The phenotypic and genotypic heterogeneities indicate that no isolated genetic event accounts for gliomagenesis, but rather that it is the cumulative effects of a number of alterations that operate in a concerted manner. In this pathological process are included activation of growth factor receptor signaling pathways, downregulation of many apoptotic mechanisms, and imbalance of pro- and antiangiogenic factors. Several growth factor receptors, such as epidermal growth factor receptor (EGFR), platelet-derived growth factor receptor (PDGFR), C-Kit, vascular endothelial growth factor receptor (VEGFR) are overexpressed, amplified, or mutated in gliomas. Hence, the modulation of gene expression at more levels, such as DNA, messenger RNA, proteins, and transduction signal pathways, may represent the most effective modality to downregulate or silence some specific gene functions in neoplastic cells.[22]

Regardless of all these research advancements, efficacious treatments to cure GBM have yet to be developed. Standard GBM treatment includes maximal surgical resection with a postoperative combination of radiotherapy with concomitant and adjuvant TMZ chemotherapy.[23] Although this treatment improves the median overall survival from 6 to 14.6 months, GBM remains a lethal tumor. Maximal safe resection of a primary GBM still remains the mainstay and confers improved prognosis. Alarmingly, with this therapy, patient survival at 5 years is below 10%. This is in part due to the invasive behavior of the tumor and the resulting inability to resect more than 98% of the bulk. For this reason, recurrence even after the most advanced treatment may be inevitable, and even in cases where apparent gross total resection is achieved.[15] Specifically, patients who receive a surgical resection greater than 98% of the tumor volume have a prognosis of 13.1 months compared with 8.8 months in patients from whom less of the tumor is resected.[24] The indefinable borders of GBM cell infiltration into the surrounding healthy tissue prevent complete surgical removal. For this reason, most GBM patients will follow a standard treatment regimen after the tumor is resected. This consists of 6 weeks of external beam radiation five times a week plus daily oral TMZ. However, TMZ is an alkylating agent that does not always have therapeutic efficacy on each tumor cell. Unfortunately, most patients will have a recurrence within 6.9 months of their primary diagnosis. Essentially, all GBMs recur, and, among these, at least 80% of GBM recurrences occur in the same area as the original tumor, but re-operation and re-radiation are options for only a minority of patients. In addition, genetic mutations, epigenetic modifications and microenvironmental heterogeneity cause resistance to radiotherapy and chemotherapy, resulting in a therapeutic scenario that is difficult to overcome. Therefore, the development of efficient therapeutic strategies to combat these tumors requires a better knowledge of genetic and proteomic alterations as well as of the infiltrative behavior of GBM cells and how this can be targeted.

Cerebral gliomas show a unique pattern of invasion and exceptionally metastasize outside the central nervous system. Their invasion comprises the translocation of active malignant cells through host cellular and extracellular matrix (ECM) barriers.[25,26] How they can evade immune detection and defer commitment to proliferation, remains mostly unknown. Basically, invading glioma cells migrate to distinct anatomical structures: the basement membrane of blood vessels, the subependymal space, the glial limitans externa, and parallel and intersecting nerve fiber tracts in the white matter. They adhere to proteins of the surrounding ECM, are able to degrade ECM components by secretion of proteases and migrate. The ECM is composed of proteoglycans, glycoproteins, and collagens and also contains fibronectin, laminin, tenascin, hyaluronic acid, and vitronectin. Key points in the process of invasion are the synthesis and deposition of ECM components by glioma and mesenchymal cells, the release of ECM-degrading factors for remodeling interstitial spaces, the presence of adhesion molecules, and the effects of cell–matrix interactions on the behavior of glioma cells. The importance of ECM modification lies in the loss of contact inhibition, allowing tumor cells to freely migrate. It is known that the proteolytic degradation of the basement membrane is mediated by proteases, such as the matrix metalloproteinases (MMPs), secreted by tumor and stromal cells.[27] Also, the brain tumor invasion is probably the result of an imbalance between the production of MMPs and of tissue inhibitor of metalloproteases-1 produced by the tumor cells.[27] MMP-1 initiates the breakdown of the interstitial collagens and activates the other MMPs to allow glioma cell infiltration. Glioma cells adhere to each other and to the ECM components through cell adhesion molecules such as integrins, selectins, cadherins, the immunoglobulin superfamily, and lymphocyte homing receptors. Integrins are heterodimers composed of α- and β-subunits. These proteins regulate many aspects of cell behavior, including survival, proliferation, migration, and differentiation. Integrins are expressed on different cell types, including neurons, glial cells, and meningeal and endothelial cells. Downregulated β_1 integrin levels seem to affect the locally invasive behavior of astrocytic tumors through the interaction of glioma cells with ECM components, leading to reduced migration along vascular basement membranes.[28]

Numerous chemotherapeutic drugs affecting cell division or DNA synthesis have been tested to improve the prognosis of GBM, including alkylating agents that differ from TMZ (Busulfan, Bicnu, Carboplatin, Carmustine, Cisplatin, Lomustine, Oxaliplatin), topoisomerase inhibitors (Irinotecan,[29] Topotecan), anthracyclines (Doxorubicin), and antimitotic agents (Vincristine, Taxanes).[30] Furthermore, new innovative anticancer drugs targeting receptor tyrosine kinases – EGFR, PDGFR, VEGFR – and their downstream signaling pathways PI3k/Akt/mTOR, Ras/mitogen-activated protein kinase, protein kinase C – have also been developed and used in clinical trials.[31] However, the greatest obstacle in the treatment of brain tumors with these drugs is often not drug potency, but the physical barrier, the blood–brain barrier, that renders the usual circulatory routes of delivery ineffective.[32]

The Cancer Genome Atlas research network performed the whole genome sequencing of GBMs and found a link between GBM initiation, progression, and recurrence and epigenetic pathways, that have recently emerged as possible new drug targets. In 90% of patients, GBMs arise de novo as primary tumors without progression from lower grades, whereas secondary GBMs originate from previously diagnosed low-grade astrocytomas. Greater knowledge of the cellular, genetic, and epigenetic origins of GBMs is the key for developing personalized GBM treatment. Recent attempts to characterize the GBM epigenetic/genetic landscape have used integrative approaches including multiple types of high-throughput data such as genetic mutations and RNA expression, as well as methylation and protein expression.[33] Novel therapies are likely to come from epigenetic targets, which offer the opportunity to modulate aberrant cellular behavior.

Examples of epigenetic therapies originate from the observation of the behavior of invasive glioma cells as a consequence of their reaction to specific components of the ECM that act as substrates for glioma spread.[34,35] Various anti-invasive techniques have been shown to inhibit the proliferative processes by in vitro and in vivo models.[36–39] The molecular pharmacotherapeutic approach to high-grade gliomas includes gene therapy, antisense oligonucleotides, immunotherapy, small molecules inhibiting tyrosine kinase, farnesyltransferase and MMPs. Among these, the antisense oligonucleotides appear the most promising. This method is based on the sequence-specific binding of an antisense oligonucleotide to target messenger RNA. Antisense oligonucleotides are synthetic stretches of DNA that hybridize with specific messenger RNA strands. The specificity of hybridization makes the antisense method able to selectively modulate the expression of genes involved in tumorigenesis, so allowing the prevention of gene translation. The goal of an antisense molecule-based approach is to selectively suppress the expression of a protein by exploiting the genetic sequence in which it is encoded. The identification and validation of antisense inhibitors is the fastest way to identify inhibitors of gene expression. Moreover, as they target RNA instead of proteins, they can be extremely specific and versatile when properly used.[40–42]

The number of potential targets for antisense oligonucleotide treatment of glioma cells is extremely large, including genes coding for growth factors and their receptors, cellular proteases, kinases, second messengers, proto-oncogenes, and factors and proteins that are important in cell cycle control and apoptosis. The classical examples of use of antisense oncogenes involve *c-myb* or *bcl-2*. In neoplasm progression, several growth factor receptors, such as EGFR, VEGFR, transforming growth factor-β, insulin-like growth factor 1 receptor are overexpressed, amplified, and mutated in gliomas.[43] The clinical results obtained using antisense therapy are often similar to those obtained by using inhibitor chemotherapeutic agents (i.e. imatinib, gefitinib) targeting the growth factors and the downstream elements of signaling pathways. The results are especially interesting when applying the combined techniques of antisense and/or inhibitors with chemotherapy. Using either the approach of inhibitors of growth factors and their receptors, especially of EGF, or the approach of antisense insulin-like growth factor 1 and anti-transforming growth factor-β inducing an antitumor response, together with TMZ chemotherapy, the median survival of GBM patients can actually reach 2 years. These results constitute progress compared with the classical treatment and they underline the value of molecular-biology-based gene therapy, especially immune-gene therapy.[44]

However, for these targets to be validated in GBM, detailed network analyses are required to be coupled to understanding the role that epigenetic enzymes play in the survival of GBM cells of origin.

Once an epigenetic target is selected, the next challenge will be to generate small molecule inhibitors that can pass through the blood–brain barrier and reach the target malignant cells. Given the likelihood that drug combinations will be required to effectively treat GBM, drug interactions will also have to be considered to identify the effect of each drug in influencing blood–brain barrier permeability.

Unfortunately, the successful therapeutic responses have not statistically met with the expectations engendered by the enormous progress in cancer molecular genetics that was made in the latter part of the twentieth century.

References

1. Ferlay, J., Shin, H.R., Bray, F., Forman, D., Mathers, C., and Parkin, D.M. (2010) Estimates of worldwide burden of cancer in 2008: Globocan 2008. *Int. J. Cancer*, **127**: 2893–917.
2. Schwartzbaum, J.A., Fisher, J.L., Aldape, K.D., and Wrensch, M. (2006) Epidemiology and molecular pathology of glioma. *Nat. Clin. Pract. Neurol.*, **2**: 494–503.
3. Central Brain Tumor Registry of the United States. Available from: *http://www.CBTRUS.org*. [Accessed 20 December 2013].
4. Wrensch, M., Minn, Y., Chew, T., Bondy, M., and Berger, M.S. (2002) Epidemiology of primary brain tumors: current concepts and review of the literature. *NeuroOncol.*, **4**: 278–99.
5. Inskip, P.D., Linet, M.S., and Heineman, E.F. (1995) Etiology of brain tumors in adults. *Epidemiol. Rev.*, **17**: 382–414.
6. Louis, D.N., Ohgaki, H., Wiestler, O.D., Cavenee, W.K., Burger, P.C., et al. (2007) The 2007 WHO classification of tumours of the central nervous system. *Acta Neuropathol.*, **114**: 97–109.
7. Tortosa, A., Viñolas, N., Villà, S., Verger, E., Gil, J.M., et al. (2003) Prognostic implication of clinical, radiologic, and pathologic features in patients with anaplastic gliomas. *Cancer*, **97**: 1063–71.

8. Compostella, A., Tosoni, A., Blatt, V., Franceschi, E., and Brandes, A.A. (2007) Prognostic factors for anaplastic astrocytomas. *J. Neurooncol.*, **81**: 295–303.
9. McGirt, M.J., Chaichana, K.L., Gathinji, M., Attenello, F.J., Than, K., et al. (2009) Independent association of extent of resection with survival in patients with malignant brain astrocytoma. *J. Neurosurg.*, **110**: 156–62.
10. Korshunov, A., Golanov, A., and Sycheva, R. (2002) Immunohistochemical markers for prognosis of anaplastic astrocytomas. *J. Neurooncol.*, **58**: 203–15.
11. Hartmann, C., Hentschel, B., Wick, W., Capper, D., Felsberg, J., et al. (2010) Patients with IDH1 wild type anaplastic astrocytomas exhibit worse prognosis than IDH1-mutated glioblastomas, and IDH1 mutation status accounts for the unfavorable prognostic effect of higher age: implications for classification of gliomas. *Acta Neuropathol.*, **120**: 707–18.
12. Smith, J.S., Tachibana, I., Passe, S.M., Huntley, B.K., Borell, T.J., et al. (2001) PTEN mutation, EGFR amplification, and outcome in patients with anaplastic astrocytoma and glioblastoma multiforme. *J. Natl. Cancer Inst.*, **93**: 1246–56.
13. Cairncross, G., Wang, M., Shaw, E., Jenkins, R., Brachman, D., et al. (2013) Phase III trial of chemoradiotherapy for anaplastic oligodendroglioma: long-term results of RTOG 9402. *J. Clin. Oncol.*, **31**: 337–43.
14. Perry, A., Jenkins, R.B., O'Fallon, J.R., Schaefer, P.L., Kimmel, D.W., et al. (1999) Clinicopathologic study of 85 similarly treated patients with anaplastic astrocytic tumors: an analysis of DNA content (ploidy), cellular proliferation, and p53 expression. *Cancer*, **86**: 672–83.

15. Clarke, J., Penas, C., Pastori, C., Komotar, R.J., Bregy, A., et al. (2013) Epigenetic pathways and glioblastoma treatment. *Epigenetics*, **8**(8): 785–95.
16. Chen, J., McKay, R.M., and Parada, L.F. (2012) Malignant glioma: lessons from genomics, mouse models, and stem cells. *Cell*, **149**: 36–47.
17. Liu, C., Sage, J.C., Miller, M.R., Verhaak, R.G., Hippenmeyer, S., et al. (2011) Mosaic analysis with double markers reveals tumor cell of origin in glioma. *Cell*, **146**: 209–21.
18. Koso, H., Takeda, H., Yew, C.C., Ward, J.M., Nariai, N., et al. (2012) Transposon mutagenesis identifies genes that transform neural stem cells into glioma-initiating cells. *Proc. Natl. Acad. Sci. USA*, **109**: E29983007.
19. Friedmann-Morvinski, D., Bushong, E.A., Ke, E., Soda, Y., Marumoto, T., et al. (2012) Dedifferentiation of neurons and astrocytes by oncogenes can induce gliomas in mice. *Science*, **338**: 1080–4.
20. Hambardzumyan, D., Cheng, Y.K., Haeno, H., Holland, E.C., and Michor, F. (2011) The probable cell of origin of NF1- and PDGF-driven glioblastomas. *PLoS One*, **6**: e24454.
21. Chen, J., Li, Y., Yu, T.S., McKay, R.M., Burns, D.K., et al. (2012) A restricted cell population propagates glioblastoma growth after chemotherapy. *Nature*, **488**: 522–6.
22. Caruso, G., Caffo, M., Raudino, G., and Tomasello, F. (2012) Introduction. In: Al-Jumaily (ed.) *Nanoparticles and Brain Tumor Treatment*, pp. 1–3. New York: ASME Press.
23. Caruso, G., Caffo, M., Raudino, G., Alafaci, C., Salpietro, F.M., and Tomasello, F. (2010) Antisense oligonucleotides as an innovative therapeutic strategy in

the treatment of high-grade gliomas *Rec. Pat. CNS Drug Discov.* 5 53–9.

24. Lacroix, M., Abi-Said, D., Fourney, D.R., Gokaslan, Z.L., Shi, W., et al. (2001) A multivariate analysis of 416 patients with glioblastoma multiforme: prognosis, extent of resection, and survival. *J. Neurosurg.*, **95**: 190–8.
25. Caffo, M., Germanò, A., Caruso, G., Meli, F., Galatioto, S., et al. (2004) An immunohistochemcial study of extracellular matrix proteins laminin, fibronectin and type IV collagen in paediatric glioblastoma multiforme. *Acta Neurochir.*, **146**: 1113–18.
26. Hanahan, D., and Folkman, J. (1996) Patterns and emerging mechanisms of the angiogenic switch during tumorigenesis. *Cell*, **86**: 353–64.
27. Uhm, J.H., Dooley, N.P., Villemure, J.G., and Yong, V.W. (1997) Mechanisms of glioma invasion: role of matrix metalloproteinases. *Can. J. Neurol. Sci.*, **24**: 3–15.
28. Gladson, C.L. (1996) Expression of integrin $\alpha_v\beta_3$ in small blood vessels of glioblastoma tumors. *J. Neuropathol. Exp. Neurol.*, **55**: 1143–9.
29. Huynh, N.T., Passirani, C., Saulnier, P., and Benoit, J.P. (2009) Lipid nanocapsules: a new platform for nanomedicine. *Int. J. Pharm.*, **379**: 201–9.
30. Muldoon, L.L., Soussain, C., Jahnke, K., Johanson, C., Siegal, T., et al. (2007) Chemotherapy delivery issues in central nervous system malignancy: a reality check. *J. Clin. Oncol.*, **25**: 2295–305.
31. Arko, L., Katsyv, I., Park, G.E., Luan, W.P., and Park, J.K. (2010) Experimental approaches for the treatment of malignant gliomas. *Pharmacol. Ther.*, **128**: 1–36.
32. Pardridge, W.M. (2005) Molecular biology of the blood–brain barrier. *Mol. Biotechnol.*, **30**: 57–70.

33. Riddick, G., and Fine, H.A. (2011) Integration and analysis of genome-scale data from gliomas. *Nat. Rev. Neurol.*, 7: 439–50.
34. Chintala, S.K., Gokaslan, Z.L., Go, Y., Sawaya, R., Nicolson, G.L., and Rao, J.S. (1996) Role of extracellular matrix proteins in regulation of human glioma cell invasion in vitro. *Clin. Exp. Metastasis*, **14**: 358–66.
35. Giese, A., and Westphal, M. (1996) Glioma invasion in the central nervous system. *Neurosurgery*, **39**: 235–52.
36. Tysnes, B.B., Larsen, L.F., Ness, G.O., Mahesparan, R., Edvardsen, K., et al. (1996) Stimulation of glioma-cell migration by laminin and inhibition by anti-α3 and anti-β1 integrin antibodies. *Int. J. Cancer*, **67**: 777–84.
37. Tonn, J.C., Kerkau, S., Hanke, A., Bouterfa, H., Mueller, J.G., et al. (1999) Effect of synthetic matrix-metalloproteinase inhibitors on invasive capacity and proliferation of human malignant gliomas in vitro. *Int. J. Cancer*, **80**: 764–72.
38. Groves, M.D., Puduvalli, V.K., Hess, K.R., Jaeckle, K.A., Peterson, P., et al. (2002) Phase II trial of temozolamide plus the matrix metalloproteinase inhibitor, marimastat, in recurrent and progressive glioblastoma multiforme. *J. Clin. Oncol.*, **20**: 1383–8.
39. Reardon, D.A., Akabani, G., Coleman, R.E., Friedman, A.H., Friedman, H.S., et al. (2002) Phase II trial of murine (131)I-labeled antitenascin monoclonal antibody 81C6 administered into surgically created resection cavities of patients with newly diagnosed malignant gliomas. *J. Clin. Oncol.*, **20**: 1389–97.
40. Jansen, B., and Zangemeister-Wittke, U. (2002) Antisense therapy for cancer – the tome of truth. *Lancet Oncol.*, **3**: 672–83.

41. Tamm, I. (2006) Antisense therapy in malignant disease: status quo and quo vadis? *Clin Sci.*, **110**: 427–42.
42. Wacheck, V., and Zangemeister-Wittke, U. (2006) Antisense molecules for targeted cancer therapy. *Oncol. Hematol.*, **59**: 65–73.
43. Caffo, M., Caruso, G., Passalacqua, M., Angileri, F.F., and Tomasello, F. (2013) Antisense oligonucleotides therapy in the treatment of cerebral gliomas: a review. *J. Genet. Syndr. Gene Ther.*, **4**: 10.
44. Caruso, G., Caffo, M., Pino, M.A., and Raudino, G. (2013) Antisense oligonucleotides therapy: a new potential tool in the treatment of cerebral gliomas. In: Atta-ur-Rahman (ed.) *Frontiers in Clinical Drug Research – CNS and Neurological Disorders*, Vol. 1, Chapter 3, pp. 69–115. Bentham eBooks. Available from: *http://benthamscience.com/ebooks/*.

2

Blood–brain barrier pathophysiology in brain tumors

DOI: 10.1533/9781908818744.17

Abstract: In this chapter we will focus on the blood–brain barrier physiological structure and subsequent pathophysiology in the presence of brain tumors. The existence of the blood–brain barrier selectively prevents the free passage of numerous molecules, especially macromolecules such as many drugs are. The tightness of the barrier is due to its non-fenestrated endothelial cells and junctional structure; the presence of transmembrane proteins sealing the intercellular cleft also plays a pivotal role in the selective passage of substances. In the presence of a pathological process, such as inflammation or brain tumors, the barrier undergoes modifications of its permeability. It is the thorough study of the functioning of the barrier in both physiological and pathological settings that can be the key to interfering with its permeability, and so allowing the passage of drugs.

Key words: blood–brain barrier, brain tumor, edema, pathophysiology, transmembrane proteins.

Published by Elsevier Limited, 2014

The efficacy of brain tumor therapies is hampered by the presence of the blood–brain barrier (BBB). This structure is a dynamic barrier that is selectively permeable to small molecules while macromolecules and many chemotherapeutic agents cannot overcome it unless it is partially destroyed. The existence of the BBB was first demonstrated in 1885 by the German pathologist Ehrlich.[1] He observed that intravenous injection of albumin-bound dyes into rats stained all body tissues except for the brain. Subsequent electron microscopy studies showed that intravenous administration of horseradish peroxidase is prevented from entering the brain parenchyma by capillary endothelial cells,[2] confirming the existence of a barrier.

The BBB is made up of an inner layer of endothelial cells that are continuously linked by tight junctions. Moreover, astrocytes and pericytes surround the endothelium thanks to their peduncles, creating an additional obstacle.[3] Physiologically, in addition to brain capillary endothelial cells, extracellular base membrane, adjoining pericytes, astrocytes, and microglia are all integral parts of the BBB supporting system. Together with surrounding neurons, these components form a complex and functional "neurovascular unit". That way, hydrophilic and high-molecular-weight substances cannot pass through to reach the neurons. Hence, the BBB protects the intracerebral milieu from dangerous external substances[4,5] and thanks to specialized carriers sited on its surface, it allows the influx/efflux of small lipophilic molecules, glucose, and some amino acids.[6] Despite advancements in pharmaceuticals and technology and the apparent clarity of the molecular structure and functioning of the BBB, overcoming it in order to efficaciously treat brain diseases remains a challenge, the majority of drugs have large molecular weights and do not readily permeate into the brain parenchyma, targeting the required cells.

The tightness of the BBB is due to the physical complexity of its non-fenestrated endothelial cells and junctional structure and the presence of transmembrane proteins sealing the intercellular cleft and playing a pivotal role in the presence of a pathological process. The transmembrane proteins constituting the tight junctions and bound to their counterparts on adjacent cells include claudin-1 and claudin-5, occludin, and junctional adhesion molecules.[7–9] At the cytoplasmic surface of cells, the claudins and occludin principally bind zonula occludens 1 (ZO1), ZO2, and ZO3, each of which anchors to the actin cytoskeleton. ZO1, ZO2, and ZO3 are guanylate cyclase proteins,[10] so are involved in intracellular signaling. This complex organization acts as a gate that can be transiently opened in response to second messengers to permit the passage of specific solutes.[11] Consequently, phosphorylation of occludin and claudins may regulate tight junction permeability.[7]

It is universally accepted that the endothelial tight junctions of the BBB are structurally defective in human glioblastoma (GBM), resulting in cerebral vasogenic edema.[12] However, the mechanisms underlying BBB breakdown are poorly understood. Some hypotheses come from studies on the transmembrane proteins. Occludin, claudin-1, and claudin-5 appear to be dysregulated in human glioma vessel tight junctions.[13,14] Specifically, the expression of the tight junction protein claudin-1 was lost in the majority of GBM microvessels, whereas claudin-5 and occludin were significantly downregulated only in hyperplastic vessels.[13] Moreover, as the presence of occludin in human astrocytomas is inversely related to the histological tumor grade,[14] it has been suggested that in these settings occludin is downregulated and phosphorylated.[14,15] Phosphorylation of occludin inhibits its interactions with ZO1, ZO2, and ZO3[16] and increases tight junction permeability.[7]

Complementary research supposed that glioma cells might actively degrade previously intact BBB tight junctions, depending on tumor grade. Previous studies found how abnormalities in the appearance of tight junctions in human gliomas correlate with increasing malignancy.[17] Recently, Schneider et al.[18] confirmed that cultured GBM primary cells showed marked breakdown of electrical resistance, whereas primary cultures derived from low-grade gliomas showed delayed or no effects. These results suggest that malignant gliomas have acquired the ability to actively degrade tight junctions by secreting soluble factors, eventually leading to BBB disruption within invaded brain tissue.[18]

Another concurrent mechanism set up to weaken the physiological functioning of the BBB address the presence of malignant astrocytes and/or pericytes having lost their ability to induce BBB features of cerebral endothelial cells due to dedifferentiation.[18] In a normal environment, astrocytes and pericytes secrete factors that confer BBB properties to adjacent endothelial cells. This has been confirmed by observing transplanted astrocytes in vivo and in vitro inducing BBB properties in adjacent non-neural endothelial cells from different species.[19,20] Cerebral endothelial cells that vascularize brain tumors probably dedifferentiate in response to signals produced by the surrounding tumor cells. Cultured endothelial cells from brain can be induced to lose features characteristic of their tissue of origin and adopt an immature phenotype by changing the culture media and exposing them to growth factors.[21,22] In this context, it is of particular interest that the majority of microvascular pericytes were negative for a-smooth muscle actin, which is a marker of differentiated pericytes, although pericytes were frequently found in electron micrographs.[13]

Reduced numbers of normal astrocytes in tumor tissue associated with excessive secretion of angiogenic factors can be another possibility explaining the augmented permeability of the BBB in the tumor setting. Because high-grade tumors are deficient in normal astrocytes, they lack the astrocyte-derived factors required for the formation of a normal barrier.[23] In addition, brain tumor cells secrete the angiogenic vascular endothelial growth factor (VEGF),[24] and human anaplastic astrocytoma and GBM have the highest production.[25,26] In cultured brain cells, VEGF causes downregulation and phosphorylation of the tight junction proteins occludin and ZO1, with a corresponding increase in permeability.[27–29] This circle is fomented by the hypoxic environment created by human GBM cells, which outgrow their blood supply: hypoxia is a potent stimulus for secretion of VEGF by the same tumor cells.[30,31]

The extracellular matrix (ECM) also plays a role in favoring the altered permeability of the BBB. Compared with normal brain tissue, GBM has an apparently expanded and defective ECM.[32,33] The involvement of glioma-derived factors on matrix metalloproteinases has been quantitatively demonstrated in a model of cocultured GBM cells and glioma-derived transforming growth factor-β_2 (TGF-β_2). The presence of TGF-β_2 enhanced the paracellular flux of endothelial cell monolayers in conjunction with significant downregulation of the tight junction proteins occludin, claudin-1, and claudin-5.[34]

All of these phenomena are restricted to the tumor site. Recent in vitro studies evaluating brain tumor-induced alterations and BBB permeability in brain capillary endothelial cells of mice indicated that, although BBB integrity is altered within the tumor site at later stages of development, the BBB is generally still functional and limiting in terms of solute and drug permeability in and around the tumor.[35]

To be able to interfere with the permeability of the BBB bypassing it to treat brain diseases, we need also to take advantage of the transport mechanisms across the barrier. Concentration differences, lipophilicity, and molecular weight drive the passage so that only small lipid-soluble molecules with a molecular weight of 400 Daltons can simply cross the BBB. Due to the unique properties of the BBB, paracellular transport of hydrophilic drugs is virtually absent and transcellular transport by passive diffusion is only available to molecules with a low molecular weight.[36–38] Only a small percentage of drugs fit these criteria.[39] For other molecules, their transport across the BBB will then have to rely on either the integrity of the BBB or the drug properties and their affinity for BBB receptors. For these other substances, transport proteins acting as carriers, or as specific receptor-mediated or vesicular mechanisms are required.

The passage mechanisms across the BBB imply not only simple diffusion, either paracellular or transcellular, but also facilitated transport (Figure 2.1).[40] Small water-soluble molecules simply diffuse through the tight junctions but not to any great extent. Small lipid soluble substances like alcohol and steroid hormones penetrate transcellularly by dissolving in their lipid plasma membrane. However, for almost all other substances transport proteins (carriers), specific receptor-mediated or vesicular mechanisms (adsorptive transcytosis) are required to pass the BBB.[41]

Facilitated (passive) diffusion needs specific membrane protein carriers to carry across the BBB a substance whose external concentration is high – i.e. carrier-mediated transport (CMT). In the case of transport proteins, or CMT, there is binding of a solute such as glucose or amino acids to a protein transporter on one side of the membrane, which triggers a conformational change in the protein, resulting in the transport of the substance to the other side of the membrane, from high to low concentration. If compounds need to be moved against a concentration gradient, ATP may

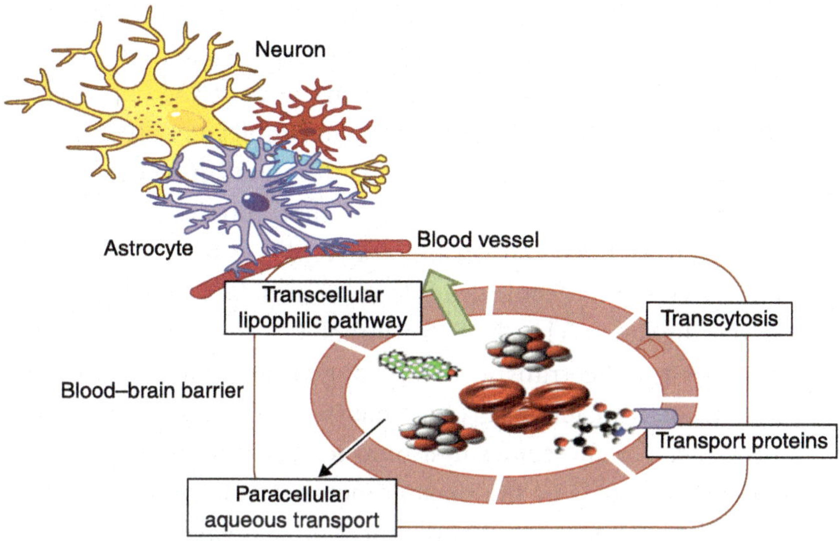

Figure 2.1 An illustration of the component of the blood–brain barrier, including the main routes for transport of substances across the barrier itself: (a) paracellular aqueous pathway; (b) transcellular lipophilic pathway; (c) transport proteins; and (d) transcytosis

provide the energy to facilitate the process. Efflux pumps or transporters are responsible for extruding drugs from the brain and this mechanism is a major obstacle for the accumulation of a wide range of biologically active molecules in the brain, with the ATP binding cassette transporter P-glycoprotein and multidrug-resistant protein being the principle efflux mechanism of these agents.[42] Inhibition of P-glycoprotein in preclinical studies has enhanced the penetration of paclitaxel into the brain, indicating the feasibility of achieving improved drug delivery to the brain by suppression of P-glycoprotein.[43] CMT systems can be exploited for brain drug delivery after reformulating the drug in such a way that the drug assumes a molecular structure that mimics that of the endogenous ligand.

Receptor-mediated transport/transcytosis (RMT) provides a means for selective uptake of macromolecules. Just as the CMT systems are portals of entry for small-molecule drugs that have a molecular structure that mimics that of an endogenous CMT substrate, the RMT systems allow the entry of large-molecule drugs that are attached to endogenous RMT ligands. The BBB endothelial cells express RMT systems for the transport of endogenous peptides, such as insulin or transferrin, growth factors, enzymes, and plasma proteins. The molecules first bind to receptors that collect in specialized areas of the plasma membrane known as coated pits. When bound to ligand these pits invaginate into the cytoplasm and then pinch free of the plasma membrane to form coated vesicles. After acidification of the endosome, the ligand will dissociate from the receptor and cross the other side of membrane.[44,45] The RMT systems operate in parallel with the classical CMT. The RMT has particular interest for drug delivery because it is mainly employed in the transport of macromolecules like peptides and proteins across the BBB, by conjugating the substance with ligands such as lactoferrin, transferrin, and insulin.

Endocytosis and transcytosis allow the internalization, sorting and trafficking of many plasma macromolecules. Through endocytosis molecules from the circulation are internalized in vesicles and are directed to endosomes or lysosomes within the cell. Receptor mediated endocytosis is clathrin-dependent and provides for a highly specific and energy-mediated transport that enables cells to selectively uptake macromolecules.

Adsorptive-mediated transcytosis (AMT) or pinocytosis facilitates the transport of large peptides such as immunoglobulin G, albumin, and native ferritin thanks to an electrostatic interaction between a positively charged substance and the negatively charged sites on the brain endothelial cell surface.[46] AMT has a lower affinity but higher capacity than RMT. The development of many new drug-delivery technologies focuses on AMT.[47] Another type of transcytosis is cell-mediated transcytosis;[48] this relies on immune cells such as monocytes or macrophages to cross an intact BBB. In fact, it is a well-established mechanism for the entry of some pathogens, such as *Cryptococcus neoformans* and human immunodeficiency virus, into the brain, and is known as the "Trojan horse" model.[49,50] Unlike other pathways of transport, which normally permit only solute molecules with specific properties, cell-mediated transcytosis can be used for almost any type of molecules or materials as well as particulate carrier systems.[51] An increased site specificity and internalization can improve the efficacy of treatments based on this kind of transport and decrease the possibility of side effects. Another relatively new target is represented by the aquaporins (AQPs), involved in the pathophysiology of brain edema.[52–54] The AQPs are a family of 12 water-channel proteins, of which AQP4 is predominant in brain; it is strongly expressed in astrocyte foot processes, and has been shown to increase the water permeability in vitro.[55,56] AQP4 appears strongly upregulated in human astrocytic tumors[57] and is also found throughout the entire astrocyte cell membrane. Indirect evidence suggests a correlation between the size of the ECM and the level of AQP4 expression, with AQP4-expressing gliotic tissue presenting a significant barrier to extracellular fluid flow.[58] It has been recently reported that AQP1, which is normally expressed in choroid plexus, can also be expressed in tumor cells and peritumoral astrocytes in human high-grade gliomas.[59,60] Its role in the pathophysiology of brain tumor edema is not known.

Endothelial tight junction proteins, transport systems across the barrier, and AQP4 may represent targets for the design of novel drugs to treat brain tumors. By using nanotechnology it is possible to deliver the drug to the targeted tissue across the BBB, release the drug at a controlled rate, and reduce peripheral toxicity.

References

1. Ehrlich, P. (1885) *Das sauerstoff-bedurnfniss des oraganismus: eine farbenanalytische studie*. Berlin: Hirschold.
2. Reese, T.S., and Karnovsky, M.J. (1967) Fine structural localization of a blood–brain barrier to exogenous peroxidase. *J. Cell. Biol.*, **34**: 207–17.
3. Egleton, R.D., and Davis, T.P. (1997) Bioavailability and transport of peptides and peptide drugs into the brain. *Peptides*, **18**: 1431–9.
4. Caruso, G., Caffo, M., Alafaci, C., Raudino, G., Cafarella, D., et al. (2011) Could nanoparticles systems have a role in the treatment of cerebral gliomas? *Nanomedicine*, **7**: 744–52.
5. Caruso, G., Raudino, G., Caffo, M., Alafaci, C., Granata, F., et al. (2010) Nanotechnology platforms in diagnosis and treatments of primary brain tumors. *Recent Pat. Nanotechnol.*, **4**: 119–22.
6. Deeken, J.F., and Löscher, W. (2007) The blood–brain barrier and cancer: transporters, treatment, and Trojan horses. *Clin. Cancer Res.*, **13**: 1663–74.
7. Rubin, L.L., and Staddon, J.M. (1999) The cell biology of the blood–brain barrier. *Annu. Rev. Neurosci.*, **22**: 11–28.

8. Wolburg, H., and Lippoldt, A. (2002) Tight junctions of the blood–brain barrier: development, composition and regulation. *Vascul. Pharmacol.*, **38**: 323–37.
9. Vorbrodt, A.W., and Dobrogowska, D.H. (2003) Molecular anatomy of intercellular junctions in brain endothelial and epithelial barriers: electron microscopist's view. *Brain Res. Brain Res. Rev.*, **42**: 221–42.
10. Haskins, J., Gu, L., Wittchen, E.S., Hibbard, J., Stevenson, B.R. (1998) ZO-3, a novel member of the MAGUK protein family found at the tight junction, interacts with ZO-1 and occludin. *J. Cell Biol.*, **141**: 199–208.
11. Vestweber, D. (2002) Regulation of endothelial cell contacts during leukocyte extravasation. *Curr. Opin. Cell Biol.*, **14**: 587–93.
12. Stewart, P.A., Hayakawa, K., Hayakawa, E., Farrell, C.L., and Del Maestro, R.F. (1985) A quantitative study of blood-brain barrier permeability ultrastructure in a new rat glioma model. *Acta Neuropathol. (Berl.)*, **67**: 96–102.
13. Liebner, S., Fischmann, A., Rascher, G., Duffner, F., Grote, E.H., et al. (2000) Claudin-1 and claudin-5 expression and tight junction morphology are altered in blood vessels of human glioblastoma multiforme. *Acta Neuropathol. (Berl.)*, **100**: 323–31.
14. Papadopoulos, M.C., Saadoun, S., Woodrow, C.J., Davies, D.C., CostaMartins, P., et al. (2001) Occludin expression in microvessels of neoplastic and non-neoplastic human brain. *Neuropathol. Appl. Neurobiol.*, **27**: 384–95.
15. Papadopoulos, M.C., Saadoun, S., Davies, D.C., and Bell B.A. (2001) Emerging molecular mechanisms of brain tumour oedema. *Br. J. Neurosurg.*, **15**: 101–8.

16. Kale, G., Naren, A.P., Sheth, P., and Rao, R.K. (2003) Tyrosine phosphorylation of occludin attenuates its interactions with ZO-1, ZO-2, and ZO-3. *Biochem. Biophys. Res. Commun.*, **302**: 324–9.
17. Shibata, S. (1989) Ultrastructure of capillary walls in human brain tumors. *Acta Neuropathol. (Berl.)*, **78**: 561–71.
18. Schneider, S.W., Ludwig, T., Tatenhorst, L., Braune, S., Oberleithner, H., et al. (2004) Glioblastoma cells release factors that disrupt blood-brain barrier features. *Acta Neuropathol.*, **107**: 272–6.
19. Janzer, R.C., and Raff, M.C. (1987) Astrocytes induce blood-brain barrier properties in endothelial cells. *Nature*, **325**: 253–7.
20. Hayashi, Y., Nomura, M., Yamagishi, S., Harada, S., Yamashita, J., and Yamamoto, H. (1997) Induction of various blood–brain barrier properties in non-neural endothelial cells by close apposition to cocultured astrocytes. *Glia*, **19**: 13–26.
21. Dropulic, B., and Masters, C.L. (1987) Culture of mouse brain capillary endothelial cell lines that express factor VIII, γ-glutamyl transpeptidase, and form junctional complexes in vitro. *In Vitro Cell Dev. Biol.*, **23**: 775–81.
22. Stolz, D.B., and Jacobson, B.S. (1991) Macro- and microvascular endothelial cells in vitro: maintenance of biochemical heterogeneity despite loss of ultrastructural characteristics. *In Vitro Cell Dev. Biol.*, **27A**: 169–82.
23. Janzer, R.C., and Raff, M.C. (1987) Astrocytes induce blood–brain barrier properties in endothelial cells. *Nature*, **325**: 253–7.
24. Bates, D.O., Lodwick, D., and Williams, B. (1999) Vascular endothelial growth factor and microvascular permeability. *Microcirculation*, **6**: 83–96.

25. Lafuente, J.V., Adan, B., Alkiza, K., Garibi, J.M., Rossi, M., and Cruz-Sanchez, F.F. (1999) Expression of vascular endothelial growth factor (VEGF) and platelet-derived growth factor receptor-β (PDGFR-β) in human gliomas. *J. Mol. Neurosci.*, **13**: 177–85.
26. Ludwig, H.C., Feiz-Erfan, I., Bockermann, V., Behnke-Mursch, J., Schallock, K., and Markakis, E. (2000) Expression of nitric oxide synthase isozymes (NOS I–III) by immunohistochemistry and DNA in situ hybridization: correlation with macrophage presence, vascular endothelial growth factor (VEGF) and oedema volumetric data in 220 glioblastomas. *Anticancer Res.*, **20**: 299–304.
27. Antonetti, D.A., Barber, A.J., Hollinger, L.A., Wolpert, E.B., and Gardner, T.W. (1999) Vascular endothelial growth factor induces rapid phosphorylation of tight junction proteins occludin and zonula occludens 1: a potential mechanism for vascular permeability in diabetic retinopathy and tumors. *J. Biol. Chem.*, **274**: 23463–7.
28. Wang, W., Dentler, W.L., and Borchardt, R.T. (2001) VEGF increases BMEC monolayer permeability by affecting occludin expression and tight junction assembly. *Am. J. Physiol. Heart Circ. Physiol.*, **280**: H434–40.
29. Behzadian, M.A., Windsor, L.J., Ghaly, N., Liou, G., Tsai, N.T., and Caldwell, R.B. (2003) VEGF-induced paracellular permeability in cultured endothelial cells involves urokinase and its receptor. *FASEB J.*, **17**: 752–4.
30. Plate, K.H., Breier, G., Weich, H.A., Mennel, H.D., and Risau, W. (1994) Vascular endothelial growth factor and glioma angiogenesis: coordinate induction of VEGF receptors, distribution of VEGF protein and

possible in vivo regulatory mechanisms. *Int. J. Cancer*, **59**: 520–9.

31. Shibata, T., Giaccia, A.J., Brown, J.M. (2000) Development of a hypoxiaresponsive vector for tumor-specific gene therapy. *Gene Ther.*, **7**: 493–8.
32. Bruehlmeier, M., Roelcke, U., Blauenstein, P., Missimer, J., Schubiger, P.A., et al. (2003) Measurement of the extracellular space in brain tumors using 76Br-bromide and PET. *J. Nucl. Med.*, **44**: 1210–18.
33. Vargova, L., Homola, A., Zamecnik, J., Tichy, M., Benes, V., and Sykova, E. (2003) Diffusion parameters of the extracellular space in human gliomas. *Glia*, **42**: 77–88.
34. Ishihara, H., Kubota, H., Lindberg, R.L., Leppert, D., Gloor, S.M., et al. (2008) Endothelial cell barrier impairment induced by glioblastomas and transforming growth factor *β*2 involves matrix metalloproteinases and tight junction proteins. *J. Neuropathol. Exp. Neurol.*, **67**: 435–48.
35. On, N.H., Mitchell, R., Savant, S.D., Bachmeier, C.J., Hatch, G.M., and Miller, D.W. (2013) Examination of blood–brain barrier (BBB) integrity in a mouse brain tumor model. *J. Neurooncol.*, **111**: 133–43.
36. Pathan, S.A., Iqbal, Z., Zaidi, S.M., Talegaonkar, S., Vohra, D., et al. (2009) CNS drug delivery systems: novel approaches. *Recent Pat. Drug Deliv. Formul.*, **3**: 71–89.
37. Chen, Y., Dalwadi, G., and Benson, H. (2004) Drug delivery across the blood–brain barrier. *Curr. Drug Deliv.*, **1**: 361–76.
38. de Boer, A.G., and Gaillard, P.J. (2007) Drug targeting to the brain. *Annu. Rev. Pharmacol. Toxicol.*, **47**: 323–55.
39. Pardridge, W.M. (2003) Blood–brain barrier drug targeting: the future of brain drug development, *Mol. Interv.*, **3**: 90–105; 51.

40. Caffo, M., Raudino, G., and Caruso, G. (2013) Nanotechnology and brain tumors drug delivery. *Recent Pat Nanomed.*, 3(1): 26–36.
41. Chen, Y., and Liu, L. (2012) Modern methods for delivery of drugs across the blood–brain barrier. *Adv Drug Deliv. Rev.*, **64**: 640–65.
42. Lee, G., Dallas, S., Hong, M., and Bendayan, R. (2001) Drug transporters in the central nervous system: brain barriers and brain parenchyma considerations. *Pharmacol. Rev.*, **53**: 569–96.
43. Kemper, E.M., Boogerd, W., Thuis, I., Beijnen, J.H., and van Tellingen, O. (2004) Modulation of the blood–brain barrier in oncology: therapeutic opportunities for the treatment of brain tumours? *Cancer Treat. Rev.*, **30**: 415–23.
44. Rip, J., Schenk, G.J., and de Boer, A.G. (2009) Differential receptor-mediated drug targeting to the diseased brain. *Expert Opin. Drug Deliv.*, **6**: 227–37.
45. Pardridge, W.M. (2005) Drug and gene targeting to the brain via blood–brain barrier receptor-mediated transport systems. *Int. Congr. Ser.*, **1277**: 49–62.
46. Kreuter, J. (2013) Mechanism of polymeric nanoparticle-based drug transport across the blood–brain barrier (BBB). *J. Microencapsul.*, **301**: 49–54.
47. Herve, F., Ghinea, N., and Scherrmann, J.M. (2008) CNS delivery via adsorptive transcytosis. *AAPS J.*, **10**: 455–72.
48. Jain, S., Mishra, V., Singh, P., Dubey, P.K., Saraf, D.K., and Vyas, S.P. (2003) RGDanchored magnetic liposomes for monocytes/neutrophils-mediated brain targeting. *Int. J. Pharm.*, **261**: 43–55.
49. Chretien, F., Lortholary, O., Kansau, I., Neuville, S., Gray, F., and Dromer, F. (2002) Pathogenesis of cerebral

Cryptococcus neoformans infection after fungemia. *J. Infect. Dis.*, **186**: 522–30.

50. Gonzalez-Scarano, F., Martin-Garcia, J. (2005) The neuropathogenesis of AIDS. *Nat. Rev. Immunol.*, **5**: 69–81.
51. Park, K. (2008) Trojan monocytes for improved drug delivery to the brain. *J. Control. Rel.*, **132**: 75.
52. Venero, J.L., Vizuete, M.L., Machado, A., and Cano, J. (2001) Aquaporins in the central nervous system. *Prog. Neurobiol.*, **63**: 321–36.
53. Papadopoulos, M.C., Krishna, S., and Verkman, A.S. (2002) Aquaporin water channels and brain edema. *Mt Sinai J. Med.*, **69**: 242–8.
54. Amiry-Moghaddam, M., and Ottersen, O.P. (2003) The molecular basis of water transport in the brain. *Nat. Rev. Neurosci.*, **4**: 991–1001.
55. Ma, T., Yang, B., Gillespie, A., Carlson, E.J., Epstein, C.J., and Verkman, A.S. (1997) Generation and phenotype of a transgenic knockout mouse lacking the mercurial-insensitive water channel aquaporin-4. *J. Clin. Invest.*, **100**: 957–62.
56. Solenov, E., Watanabe, H., Manley, G.T., and Verkman, A.S. (2004) Sevenfold reduced osmotic water permeability in primary astrocyte cultures from AQP-4-deficient mice, measured by a fluorescence quenching method. *Am. J. Physiol. Cell Physiol.*, **286**: C426–32.
57. Saadoun, S., Papadopoulos, M.C., Davies, D.C., Krishna, S., and Bell, B.A. (2002) Aquaporin-4 expression is increased in oedematous human brain tumours. *J. Neurol. Neurosurg. Psychiatry*, **72**: 262–5.
58. Roitbak, T., and Sykova, E. (1999) Diffusion barriers evoked in the rat cortex by reactive astrogliosis. *Glia*, **28**: 40–8.

59. Saadoun, S., Papadopoulos, M.C., Davies, D.C., Bell, B.A., and Krishna, S. (2002) Increased aquaporin 1 water channel expression in human brain tumours. *Br. J. Cancer*, **87**: 621–3.
60. Oshio, K., Binder, D.K., Bollen, A., Verkman, A.S., Berger, M.S., and Manley, G.T. (2003) Aquaporin-1 expression in human glial tumors suggests a potential novel therapeutic target for tumor-associated edema. *Acta Neurochir. Suppl.*, **86**: 499–502.

3

Brain drug-delivery attempts

DOI: 10.1533/9781908818744.35

Abstract: This chapter discusses the possibilities of therapeutic overcoming of the blood–brain barrier (BBB) by different treatment strategies. Drugs are usually administered via the intravenous route, but the BBB often impedes the intravenously delivered drug in reaching the intracranial compartment. Moreover, the scarce accumulation of drugs in the brain depends also on its rapid clearance by the extracellular fluid and the plasmatic half-life of the drug itself. Nowadays, multiple strategies exist to circumvent the BBB and achieve therapeutic concentrations of drugs in the brain. The intravenous route, for example, shows more efficacy when the drug is linked to a carrier system like polymeric depots, liposomes or lipid carriers. To favor the penetration of drugs inside the brain transporter-independent and transporter-dependent mechanisms have been developed. Genetic engineering is also used to produce monoclonal antibodies and liposomes too are studied for gene therapy.

Key words: blood–brain barrier, drug delivery, genomics, monoclonal antibodies, polymer depots.

Drug delivery is the process of releasing a bioactive agent at a specific rate and at a specific site but in the current scenario

Published by Elsevier Limited, 2014

a safe and effective targeted drug-delivery system must be versatile because of the presence of thousands of new molecules. Hence, delivering a drug to its specific site of therapeutic action is one of the main actual limitations of pharmaceutical and biotechnology industries.

Drugs are usually administered via the intravenous route but the blood–brain barrier (BBB) often impedes the movement of intravenously delivered drugs into the intracranial compartment because it expresses high levels of drug efflux pumps, such as P-glycoprotein, which actively remove chemotherapeutic drugs from the brain.[1] In addition, diffusion in the brain parenchyma is very weak, which is partially due to the high intercellular fluid pressure in tumors.[2,3] Finally, brain tissue is highly sensitive, so only limited doses of therapeutic agents can be used. Current estimates are that all large-molecule drugs and 98% of small-molecule drugs minimally cross the BBB, except by leakage in areas of BBB dysfunction.[4] Moreover, it is not only the presence of the barrier that is a problem; the poor accumulation of drugs in the brain depends also on their rapid clearance by the extracellular fluid and the plasmatic half-life of the drugs themselves.[5]

Increasing experimental evidence indicates that endothelial cells and microvascular properties can be altered by the activation of neurons within the neurovascular unit that serve the specific areas.[6,7] Changes in microvascular permeability can affect astrocyte function and neuron integrity. The integration of neuron and microvascular function is a practical framework for considering the traffic of agents that might affect or improve neuron function under conditions of injury and inflammation. Neuron function could affect the window for delivery of agents through alterations in endothelial cell–astrocyte communication.

Nowadays, multiple strategies exist to circumvent the BBB and achieve therapeutic concentration of small- and large-molecular-weight drugs in the brain. The intravenous route, for example, shows more efficacy when the drug is linked to a carrier system like polymeric depots, liposomes or lipid carriers. To favor the penetration of drugs inside the brain, transporter-independent and transporter-dependent mechanisms have been developed.

Among the transporter-independent mechanisms, intranasal delivery has been used with great expectations. It is a non-invasive method of bypassing the BBB by delivering the substances to the nasal epithelium and subsequently to the olfactory cerebrospinal fluid.[8,9] The respiratory region of the nose is considered to be the major site for drug absorption into the systemic circulation, which can be by transcellular pathways or paracellular passive absorption, carrier-mediated transport, and transcytosis. The highly permeable nasal epithelium allows rapid drug absorption to the brain because of high total blood flow, porous endothelial membrane, large surface area, and avoidance of first-pass metabolism.[10] Nevertheless, while the olfactory surface area of the nasal epithelium in rodents is large, it is small in humans (50 versus 5%),[11] so intranasal delivery is not expected to achieve therapeutic drug levels in most human brain regions in vivo. Moreover, intranasal delivery can damage the nasal mucosa when used frequently, the drug can be rapidly cleared by the mucociliary system, nasal congestion can interfere with the absorption, and some of the drug may be absorbed systemically.[12]

A drug can be directly delivered inside the central nervous system (CNS) by the use of chemical substances, or by the application of energy, like ultrasonic waves or electromagnetic radiation, which help in opening the tight junctions. Because the tight junctions act as a tightly bound wall, one method for the penetration of the BBB is by disrupting it. In this case, the disruption of the BBB should be transient and reversible. Different substances have been tested to reach this goal. For example, hypertonic solutions can open the tight junctions thanks to their higher osmotic pressure, which leads to a shrinking of cerebrovascular endothelial cells and subsequent disarrangement of extracellular proteins. This way, the drug is absorbed paracellularly.[10] Effective and safe transient osmotic disruption of the BBB and the blood–tumor barrier can be achieved by intra-arterial infusion of another hyperosmotic agent, mannitol.[13] The BBB is opened to drugs, proteins, and nanoparticles for between 15 minutes (for virus-sized agents) and 4 hours (for low-molecular-weight compounds) before returning to baseline permeability.[14] This technique is currently used clinically for the delivery of chemotherapy to the CNS in patients with brain tumors. It has shown increases in parenchymal and cerebrospinal fluid chemotherapy concentrations between tenfold and 100-fold compared with intravenous administration alone. Over the past 20 years, 5645 procedures have been performed in 482 patients by institutions affiliated with the BBB Consortium with minimal adverse side effects. Significant prolongation of survival has been documented in patients with chemoresponsive tumors, such as primary CNS lymphoma, without radiotherapy and without cognitive loss.[15]

Biologically vasoactive agents such as bradykinin, angiotensin peptides, leukotrienes, histamine, and vascular endothelial growth factor are also capable of disrupting the BBB, suggesting that they may play a role in modulating BBB permeability. In the case of brain gliomas, microvascular permeability in tumor tissue is more sensitive to the effects of these compounds than the normal brain endothelial cells, thanks to a partially altered barrier. Therefore, these stimuli, when used in combination with imaging contrast agents, genes or anticancer drugs, can potentially boost the preferential delivery of these materials to the brain tumor to obtain diagnosis, chemotherapeutic treatment or gene therapy. Specifically, bradykinin, an endogenous peptide mediator of the inflammatory response, can induce transient increases in blood vessel permeability that can be highly specific for tumor vasculature. Lobadimil (RMP-7) is a synthetic bradykinin analog that is specific for the B2 receptor and is 100-fold more potent than bradykinin in mice. Pharmacological manipulation of the B2 receptor offers the possibility of highly specific opening and targeted drug delivery to tumors, albeit with the possibility that increases in delivery may only be modest and dependent on the tumor type or model treated. Clinical studies in the past 5 years have demonstrated the safety of concurrent RMP-7 and carboplatin, with or without radiation therapy, for both adults and children with gliomas.[16] It seems that RMP-7 had no effect on the pharmacokinetics or toxicity of carboplatin, and two studies have shown no objective responses of RMP-7 and carboplatin in brain stem glioma or high-grade glioma.[17] Higher doses of RMP-7 may be required to increase carboplatin delivery to tumor, but may also result in increased toxicity in normal brain.

The capacity of energy-based physical methods, such as ultrasound, microwaves, or electromagnetic fields, to open the BBB has also been investigated. The important advantage of this approach is its specificity for targeting to a specific area of the brain. Focused ultrasound techniques concentrate the energy in a deep focal area with minimal effects to surrounding tissues. This allows it to noninvasively induce targeted local biological effects. Histologically, the low-power ultrasound caused reversible focal opening that was completely healed within 24 hours. Marker dye extravasation was associated with widening of the tight junctions and active vacuole transport across the endothelial cells. Hynynen et al.[18] showed that the intravenous injection of preformed gas microbubbles before pulsed ultrasound exposure would allow localized transient opening of the BBB without causing acute damage to the neurons, apoptosis, ischemia, or long-term vascular damage. Ultrasound BBB disruption produced clinically relevant levels of liposomal doxorubicin and monoclonal antibodies in the targeted local areas of the brain in animals. It is still unclear whether this technique will show any therapeutic promise in humans. Another field that takes advantage of this method is imaging. BBB disruption by magnetic resonance imaging-guided focused ultrasound achieves focal CNS delivery in animal models.[19] In combination with an imaging device such as magnetic resonance imaging, ultrasound becomes a noninvasive approach to open targeted regions of the BBB to enable delivery of drugs and other therapeutic molecules, or of contrast agents to help in diagnosis.

A drug can also be directly injected into the brain. Intracerebral delivery involves intrathecal or intraventricular catheter strategies or controlled release matrices. These techniques are highly invasive and related to important disadvantages such as infections, catheter obstruction, and limited volume of drug distribution. To avoid open-surgery, microparticles can be easily implanted stereotaxically in precise areas of the brain without damaging the surrounding tissue. Moreover, their implantation avoids the inconvenient surgical insertion of large implants and can be repeated if required. Local and site-specific delivery of chemotherapeutic agents increases drug concentration at the tumor target, decreases systemic exposure and toxicities, and increases the duration of exposure of the tumor to the drug. Experimental and clinical studies have demonstrated a statistically significant increase in survival associated with local therapy for brain tumors.[20] Implants are made up of biodegradable and nonbiodegradable polymeric materials encapsulating drugs inside them. The basic mechanism behind drug release from these devices is diffusion.

Polymer depots have been used for the delivery of drugs into the brain tumor cavity, with the drug inside the polymer matrix as a core material. They offer sustained release of drugs thanks to the biodegradation of the polymer. Similarly, Gliadel® wafer (MGI Pharma, Bloomington, MN, USA), a polymer depot containing the drug carmustine, shows release over a period of 5 days when placed in the tumor resection cavity.[21] The therapeutic effect of Gliadel wafers was tested in two phase III clinical trials.[22] The median survival time obtained with treated patients was significantly prolonged in comparison with untreated patients (by 2 months). However, infections and cerebral edema due to the high concentration of carmustine, and obstructive hydrocephalus resulting from the dislodgment of the wafer were also present.

Physicochemical properties of drugs, such as lipophilicity, lack of ionization at physiological pH, and molecular weight, determine the extent to which drugs can cross the barrier. Chemically modifying the drugs can aid in overcoming the BBB. Lipidization – the addition of lipid-like molecules through modification of the hydrophilic moieties on the drug structure – takes advantage of the relative permeability of the BBB to lipid-soluble molecules. In this way, the drugs cross the barrier through small pores that transiently form within the lipid bilayer. The addition of hydrophobic groups to a molecule may enhance its brain transfer by passive diffusion. This strategy has been frequently employed with controversial results. Multiple chemical or enzymatic transformations before release of the active drug are employed in chemical delivery systems.[5] These chemical delivery systems are based on a dihydropyridone–quaternary pyridinium ion redox system, which relies on chemistry analogous to the ubiquitous NAD1–NADH coenzyme system. The drug is converted into a 1,4-dihydropyridine moiety-containing conjugate. After systemic administration in animals, there is extensive tissue distribution with the conjugate accessing most of the compartments, including the brain. Aiming at preventing glioma recurrence, Menei et al.[23] proposed poly(D,L-lactidecoglycolide biodegradable microspheres for the local sustained delivery of 5-fluorouracil in glioma patients.[23] The authors showed an optimal radiosensitization effect and sustained in vivo concentrations of 5-fluorouracil in the cerebrospinal fluid 1 month after microsphere implantation.[23] In a multicenter phase II trial, the effect of perioperative implantation of 5-fluorouracil-releasing microspheres followed by early radiotherapy was compared with early radiotherapy alone in patients with gross total resection of high-grade gliomas. The authors showed that implantation of 5-fluorouracil-loaded microspheres in the wall of the cavity resection increased overall survival.[24] Prodrugs are pharmacologically inactive compounds that result from transient chemical modifications of biologically active species. The chemical change is usually designed to improve some deficient physicochemical property, such as membrane permeability or water solubility. After administration, the prodrug is brought closer to the receptor site and is maintained there for longer periods. Here it is converted to the active form, usually via a single activating step. Once in the CNS, hydrolysis of the modifying group will release the active compound. Although increased lipophilicity may improve movement across the BBB, it also tends to increase uptake into other tissues, causing an increased tissue burden. This selectivity in delivery is especially detrimental when potent drugs such as steroids or cytotoxic agents are considered, because toxicity is exacerbated at nontarget sites. Moreover, while increased lipophilicity may facilitate drug uptake into the CNS, it also enhances efflux processes. This can result in poor tissue retention and short biological action.

Genetic engineering is used to produce either chimeric or humanized forms of monoclonal antibodies.[25,26] The most potent antibody-based molecular "Trojan horse" known to date is the one against the human insulin receptor.[27] Recently, this antibody has been humanized, and shown to cross the BBB in vivo in nonhuman primates.[26] Certain peptidomimetic monoclonal antibodies act as ligands for the receptor-mediated transport (RMT) systems present on the BBB. These RMT-specific antibodies bind epitopes on the receptor that are spatially removed from the endogenous ligand-binding site. The peptidomimetic monoclonal antibodies act as Trojan horses to ferry across the barrier an attached drug, protein, antisense agent, or nonviral plasmid DNA.[28–31] A number of nonantibody delivery systems have been evaluated, including histone,[32] p97,[33] receptor-associated protein,[34] the tat transduction domain peptide,[35] and other cationic peptides or polymers. Whereas the transport of ligands such as receptor-associated protein is hypothesized to be receptor-mediated, transport of cationic peptides is believed to be mediated by absorptive-mediated endocytosis systems that are based on charge interactions.[32] Delivery of biopharmaceuticals across the BBB has been reported recently using a related RMT system.[36,37] A carrier protein known as CRM197 was used as a safe and effective carrier protein in human vaccines and more recently in anticancer trials.[38] CRM197 uses the membrane-bound precursor of heparin-binding epidermal growth factor (HB-EGF) as its transport receptor. Membrane-bound HB-EGF is constitutively expressed on various tissues and cells such as BBB endothelial cells, so the brain and T lymphocytes, monocytes, and macrophages can be reached. Moreover, HB-EGF expression is upregulated strongly under inflammatory disease conditions, which will enhance targeted delivery considerably. CRM197 can deliver small interfering RNAs across the barrier by this mechanism.[37]

Trojan horse liposomes are studied for CNS gene therapy. Gene delivery across the BBB may be ineffective owing to the rapid degradation of extracellular nucleic acids, as well as the pro-inflammatory effects of naked DNA.[39] Encapsulation of plasmid DNA inside pegylated liposomes eliminates the nuclease sensitivity and proinflammatory effects of the nucleic acid.[40] Pegylated liposomes, per se, are not transported across the BBB.[41] However, the attachment of a molecular Trojan horse to the tips of the polyethylene glycol strands allows the liposome to engage the BBB RMT system, and this triggers transport of the pegylated immunoliposomes, also called Trojan horse liposomes, across the BBB.[28,42] The administration of this new technology to mice, rats, or monkeys is followed 24–48 hours later by global expression of the nonviral transgene in brain.[28,42] The intravenous injection of immunoliposomes carrying an expression plasmid encoding a short hairpin RNA directed against the human epidermal growth factor led to a 90% increase in survival time of mice with intracranial human brain cancer. The pegylated immunoliposome gene transfer technology enables intravenous RNA interference of the brain.

Other important advances that may impact brain drug delivery involve BBB genomics. This is the application of gene microarray technologies to the brain microvasculature.[43] The endothelial cells occupy a very small volume of the brain, about 0.1%. The sensitivity of gene microarray is about 10^{-4} parts. Therefore, most BBB-specific transcripts may not be detected in a whole brain gene microarray. BBB genomics starts with the isolation of RNA from the brain microvasculature. Subsequently, different technologies, such as suppressive subtractive hybridization,[44] or serial analysis of gene expression,[45] can be employed to identify those genes that are selectively expressed in brain microvasculature. BBB genomics technologies can lead to new insights into the role played by the microvasculature in brain pathology. Moreover, BBB genomics can also lead to the identification of new BBB transporters, which can then be developed as new conduits to the brain for drug targeting. In parallel with BBB genomics, BBB proteomics programs aim to use protein-based technologies to identify, at the protein level, novel targets within the BBB.[46,47]

Inhibitors of BBB active efflux transporters, such as P-glycoprotein inhibitors, have been developed.[48] Such inhibitors may act as codrugs to increase the brain penetration of P-glycoprotein substrates. This is exemplified in the case of the increased brain penetration of the chemotherapeutic agent, paclitaxel (Taxol®), by coadministration of the P-glycoprotein inhibitor, PSC-833 (valspodar).

Convection-enhanced drug delivery (CED) involves a local microinfusion of drug targeted directly to brain tissue. The basics comprise a continuous infusion pressure gradient over hours to days resulting in the distribution of therapeutic agents into the interstitial space. The CED technique is used primarily for large-molecular-weight agents that show minimal leakage across the BBB and/or have significant systemic toxicity, including viruses, oligonucleotides, nanoparticles, liposomes, and targeted immunotoxins.[49] Parameters that affect CED volume of distribution include infusion parameters (rate, volume, duration, cannula size), infusate characteristics (molecular weight, surface properties, tissue affinity), and tissue properties (tissue density, extracellular space, vascularity, and interstitial fluid pressure).[50] Animal studies have demonstrated that the volume of distribution achieved by CED can be imaged by magnetic resonance in real time by including contrast agents within the infusate.[51] The major clinical use of CED will be for targeted therapy of glioblastoma.[51] Recent studies have included interleukin-13/*Pseudomonas* exotoxin alone or in combination with radiation/temozolomide, and radioimmunotherapy with monoclonal antibodies targeting tenascin or tumor necrosis factor.[52,53] Despite promising early results, it appears that two industry-sponsored phase III trials of CED immunotoxins have been negative. Mechanisms for CED treatment failure include distribution inhomogeneity, high interstitial fluid pressure, and rapid efflux of agent from the injection site.[2] To overcome these issues, increased residence time must be achieved to enhance targeted toxin receptor binding and uptake by the cancerous cells.

References

1. Golden, P.L., and Pollack, G.M. (2003) Blood–brain barrier efflux transport. *J. Pharm. Sci.*, **92**: 1739–53.
2. Ali, M.J., Navalitloha, Y., Vavra, M.W., Kang, E.W., Itskovich, A.C., et al. (2006) Isolation of drug delivery from drug effect: problems of optimizing drug delivery parameters. *NeuroOncol.*, **8**: 109–18.
3. Navalitloha, Y., Schwartz, E.S., Groothuis, E.N., Allen, C.V., Levy, R.M., and Groothuis, D.R. (2006) Therapeutic implications of tumor interstitial fluid pressure in subcutaneous RG-2 tumors. *NeuroOncol.*, **8**: 227–33.
4. Pardridge, W.M. (2003) Molecular biology of the blood-brain barrier. *Methods Mol. Med.*, **89**: 385–99.
5. Caruso, G., Caffo, M., Raudino, G., and Tomasello, F. (2012) Blood–brain barrier. In: Al-Jumaily (ed.)

Nanoparticles and brain tumor treatment, pp. 9–13. New York: ASME Press.

6. Zhao, Y., and Wang, S. (2010) Human NT2 neural precursor-derived tumor-infiltrating cells as delivery vehicles for treatment of glioblastoma. *Hum. Gene Ther.*, **21**: 683–94.
7. Ito, S., Natsume, A., Shimato, S., Ohno, M., Kato, T., et al. (2010) Human neural stem cells transduced with IFN-β and cytosine deaminase genes intensify bystander effect in experimental glioma. *Cancer Gene Ther.*, **17**: 299–306.
8. Cavalli, R., Caputo, O., and Gasco, M.R. (2000) Preparation and characterization of solid lipid nanospheres containing paclitaxel. *Eur. J. Pharm. Sci.*, **10**: 305–9.
9. Beduneau, A., Saulnier, P., Anton, N., Hindre, F., Passirani, C., et al. (2006) Pegylated nanocapsules produced by an organic solvent-free method: evaluation of their stealth properties. *Pharm. Res.*, **23**: 2190–9.
10. Pardridge, W.M. (2007) Blood–brain barrier delivery. *Drug Discov. Today*, **12**: 54–61.
11. Klibanov, A.L., Maruyama, K., Torchilin, V.P., and Huang, L. (1990) Amphipathic polyethyleneglycols effectively prolong the circulation time of liposomes. *FEBS Lett.*, **268**: 235–7.
12. Ali, J., Ali, M., Baboota, S., Sahani, J.K., Ramassamy, C., et al. (2010) Potential of nanoparticulate drug delivery systems by intranasal administration. *Curr. Pharm. Res.*, **16**: 1644–53.
13. Gelperina, S., Maksimenko, O., Khalansky, A., Vanchugova, L., Shipulo, E., et al. (2010) Drug delivery to the brain using surfactant-coated poly(lactide-co-glycolide) nanoparticles: influence of the formulation parameters. *Eur. J. Pharm. Biopharm.*, **74**: 157–63.

14. Wang, C.X., Huang, L.S., Hou, L.B., Jiang, L., Yan, Z.T., et al. (2009) Antitumor effects of polysorbate-80 coated gemcitabine polybutylcyanoacrylate nanoparticles in vitro and its pharmacodynamics in vivo on C6 glioma cells of a brain tumor model. *Brain Res.*, **1261**: 91–9.
15. Bernardi, A., Braganhol, E., Jager, E., Figueiro, F., Edelweiss, M.I., et al. (2009) Indomethacin-loaded nanocapsules treatment reduces in vivo glioblastoma growth in a rat glioma model. *Cancer Lett.*, **281**: 53–63.
16. Hugger, E.D., Audus, K.L., and Borchardt, R.T. (2002) Effects of poly(ethylene glycol) on efflux transporter activity in Caco-2 cell monolayers. *J. Pharm. Sci.*, **91**: 1980–90.
17. Nicholson, C., and Sykova, E. (1998) Extracellular space structure revealed by diffusion analysis. *Trends Neurosci.*, **21**: 207–15.
18. Hynynen, K., McDannold, N., Vykhodtseva, N., and Jolesz, F.A. (2001) Noninvasive MR imaging-guided focal opening of the blood-brain barrier in rabbits. *Radiology*, **220**: 640–6.
19. Bobo, R.H., Laske, D.W., Akbasak, A., Morrison, P.F., Dedrick, R.L., and Oldfield, E.H. (1994) Convection-enhanced delivery of macromolecules in the brain. *Proc. Natl. Acad. Sci. USA*, **91**: 2076–80.
20. Pandit, J.K., Singh, S., and Muthu, M.S. (2006) Controlled release formulations in neurology practice. *Ann. Indian Acad. Neurol.*, **9**: 207–16.
21. Attenello, F.J., Mukherjee, D., Datoo, G., McGirt, M.J., Bohan, E., et al. (2008) Use of Gliadel (BCNU) wafer in the surgical treatment of malignant glioma: A 10-year institutional experience. *Ann. Surg. Oncol.*, **15**: 2887–93.

22. Westphal, M., Hilt, D.C., Bortey, E., Delavault, P., Olivares, R., et al. (2003) A phase 3 trial of local chemotherapy with biodegradable carmustine (BCNU) wafers (Gliadel wafers) in patients with primary malignant gliomas. *NeuroOncol.*, **5**: 79–88.
23. Menei, P., Venier, M.C., Gamelin, E., Saint-André, J.P., Hayek, G., et al. (1999) Local and sustained delivery of 5-fluorouracil from biodegradable microspheres for the radiosensitization of glioblastoma: a pilot study. *Cancer*, **86**: 325–30.
24. Menei, P., Capelle, L., Guyotat, J., Fuentes, S., Assaker, R., et al. (2005) Local and sustained delivery of 5-fluorouracil from biodegradable microspheres for the radiosensitization of malignant glioma: a randomized phase II trial. *Neurosurgery*, **56**: 242–8.
25. Coloma, M.J., Lee, H.J., Kurihara, A., Landaw, E.M., Boado, R.J., et al. (2000) Transport across the primate blood–brain barrier of a genetically engineered chimeric monoclonal antibody to the human insulin receptor. *Pharm. Res.*, **17**: 266–74.
26. Boado, R.J., Zhang, Y., Zhang, Y., and Pardridge, W.M. (2007) Humanization of anti-human insulin receptor antibody for drug targeting across the human blood–brain barrier. *Biotechnol. Bioeng.*, **96**: 381–91.
27. Pardridge, W.M., Kang, Y.S., Buciak, J.L., and Yang, J. (1995) Human insulin receptor monoclonal antibody undergoes high affinity binding to human brain capillaries in vitro and rapid transcytosis through the blood–brain barrier in vivo in the primate. *Pharm. Res.*, **12**: 807–16.
28. Shi, N., Zhang, Y., Zhu, C., Boado, R.J., and Pardridge, W.M. (2001) Brain-specific expression of an exogenous gene after i.v. administration. *Proc. Natl. Acad. Sci. USA*, **98**: 12754–9.

29. Zhang, Y., and Pardridge, W.M. (2001) Neuroprotection in transient focal brain ischemia after delayed intravenous administration of brain-derived neurotrophic factor conjugated to a blood–brain barrier drug targeting system. *Stroke*, **32**: 1378–84.
30. Wu, D., Yang, J., and Pardridge, W.M. (1997) Drug targeting of a peptide radiopharmaceutical through the primate blood–brain barrier in vivo with a monoclonal antibody to the human insulin receptor. *J. Clin. Invest.*, **100**: 1804–12.
31. Boado, R.J., Zhang, Y., Zhang, Y., Xia, C.F., and Pardridge, W.M. (2007) Fusion antibody for Alzheimer's disease with bidirectional transport across the blood-brain barrier and a β fibril disaggregation. *Bioconjug. Chem.*, **18**: 447–55.
32. Pardridge, W.M., Triguero, D., and Buciak, J. (1989) Transport of histone through the blood–brain barrier. *J. Pharmacol. Exp. Ther.*, **251**: 821–6.
33. Demeule, M., Poirier, J., Jodoin, J., Bertrand, Y., Desrosiers, R.R., et al. (2002) High transcytosis of melanotransferrin (P97) across the blood-brain barrier. *J. Neurochem.*, **83**: 924–33.
34. Prince, W.S., McCormick, L.M., Wendt, D.J., Fitzpatrick, P.A., Schwartz, K.L., et al. (2004) Lipoprotein receptor binding, cellular uptake, and lysosomal delivery of fusions between the receptor-associated protein (RAP) and a-L-iduronidase or acid a-glucosidase. *J. Biol. Chem.*, **279**: 35037–46.
35. Schwarze, S.R., Ho, A., Vocero-Akbani, A., and Dowdy, S.F. (1999) In vivo protein transduction: delivery of a biologically active protein into the mouse. *Science*, **285**: 1569–72.
36. de Boer, A.G., and Gaillard, P.J. (2007) Drug targeting to the brain. *Annu. Rev. Pharmacol. Toxicol.*, **47**: 323–55.

37. Gaillard, P.J., Brink, A., and de Boer, A.G. (2005) Diphtheria toxin receptor-targeted brain drug delivery. *Int. Cong. Series*, **1277**: 185–98.
38. Buzzi, S., Rubboli, D., Buzzi, G., Buzzi, A.M., Morisi, C., and Pironi, F. (2004) CRM197 (nontoxic diphtheria toxin): effects on advanced cancer patients. *Cancer Immunol. Immunother.*, **53**: 1041–8.
39. Norman, J., Denham, W., Denham, D., Yang, J., Carter, G., et al. (2000) Liposome-mediated, nonviral gene transfer induces a systemic inflammatory response which can exacerbate pre-existing inflammation. *Gene Ther.*, **7**: 1425–30.
40. Zhang, Y.F., Boado, R.J., and Pardridge, W.M. (2003) Absence of toxicity of chronic weekly intravenous gene therapy with pegylated immunoliposomes. *Pharm. Res.*, **20**: 1779–85.
41. Huwyler, J., Wu, D., and Pardridge, W.M. (1996) Brain drug delivery of small molecules using immunoliposomes. *Proc. Natl. Acad. Sci. USA*, **93**: 14164–9.
42. Zhang, Y., Schlachetzki, F., and Pardridge, W.M. (2003) Global non-viral gene transfer to the primate brain following intravenous administration. *Mol. Ther.*, **7**: 11–18.
43. Pardridge, W.M. (2007) Blood–brain barrier genomics. *Stroke*, **38**(2 Suppl.): 686–90.
44. Li, J.Y., Boado, R.J., and Pardridge, W.M. (2001) Blood–brain barrier genomics. *J. Cereb. Blood Flow Metab.*, **21**: 61–8.
45. Enerson, B.E., and Drewes, L.R. (2006) The rat blood–brain barrier transcriptome. *J. Cereb. Blood Flow Metab.*, **26**: 959–73.
46. Shusta, E.V. (2005) Blood–brain barrier genomics, proteomics, and new transporter discovery. *NeuroRx*, **2**: 151–61.

47. Shusta, E.V., Boado, R.J., and Pardridge, W.M. (2002) Vascular proteomics and subtractive antibody expression cloning. *Mol. Cell Proteomics*, **1**: 75–82.
48. Hoffmann, K., and Loscher, W. (2007) Upregulation of brain expression of P-glycoprotein in MRP2-deficient TR(–) rats resembles seizure-induced up-regulation of this drug efflux transporter in normal rats. *Epilepsia*, **48**: 631–45.
49. Vogelbaum, M.A. (2005) Convection enhanced delivery for the treatment of malignant gliomas: symposium review. *J. Neurooncol.*, **73**: 57–69.
50. Raghavan, R., Brady, M.L., Rodriguez-Ponce, M.I., Hartlep, A., Pedain, C., and Sampson, J.H. (2006) Convection-enhanced delivery of therapeutics for brain disease, and its optimization. *Neurosurg. Focus*, **20**: E12.
51. Krauze, M.T., McKnight, T.R., Yamashita, Y., Bringas, J., Noble, C.O., et al. (2005) Real-time visualization and characterization of liposomal delivery into the monkey brain by magnetic resonance imaging. *Brain Res. Brain Res. Protoc.*, **16**: 20–6.
52. Kunwar, S., Prados, M.D., Chang, S.M., Berger, M.S., Lang, F.F., et al. (2007) Direct intracerebral delivery of cintredekin besudotox (IL13-PE38QQR) in recurrent malignant glioma: a report by the Cintredekin Besudotox Intraparenchymal Study Group. *J. Clin. Oncol.*, **25**: 837–44.
53. Yang, W., Barth, R.F., Wu, G., Kawabata, S., Sferra, T.J., et al. (2006) Molecular targeting and treatment of EGFRvIII-positive gliomas using boronated monoclonal antibody L8A4. *Clin. Cancer Res.*, **12**: 3792–802.

4

Nanoparticles potential: types, mechanisms of action, actual in vitro and animal studies, recent patents

DOI: 10.1533/9781908818744.53

Abstract: The notion of nanotechnology has evolved since its futuristic conception to its current position as a mainstream research initiative with broad applications among all divisions of science. One of the most important clinical applications of nanotechnology will undoubtedly be realized in therapeutics, and the challenge of brain tumor treatment can greatly benefit from it. Their nanometric size along with the electrostatic charge and putative lipophilic characteristics allow nanoparticles to penetrate into the brain tissue freely. Nanoparticles across the blood–brain barrier can represent ideal devices for bypassing that obstacle, becoming the new frontier for successful brain tumor therapy. Intravenous nanoparticles are administered, in experimental settings, as carriers to deliver drugs into the tumor bed. Although promising in vitro results have been reported, it remains unclear how effective such a system would be due to intra- and inter-individual patient heterogeneity. Relevant patents of nanoparticle systems used as drug-delivery carriers are also reported.

Published by Elsevier Limited, 2014

Key words: brain tumor, nanomedicine, nanoparticles, nanotechnology, patent.

Current advances in biotechnology and pharmacology need the contemporary discovery and rational design of new classes of drugs, so it is crucial to improve specific drug-delivery methods to turn these advances into clinical effectiveness. Most of the drugs possess limited solubility, high toxicity, or need high dosage or aggregation because of poor solubility, have a nonspecific delivery, are degraded in vivo or have short circulating half-lives. Targeted drug-delivery systems can convey drugs more effectively and conveniently, increase patient compliance, and reduce healthcare costs.[1] In addition, novel drug-delivery systems would be safer and improve the pharmacokinetics of easily degradable peptides and proteins that often have short half-lives in vivo.[2,3] Therefore, the development of techniques that could selectively deliver drugs to the pathological sites is currently one of the most important areas of drug research. The emergence of nanotechnology is likely to have a significant impact on the drug-delivery sector and nanoparticles (NPs) are at the leading edge, with many potential applications in clinical medicine and research.

4.1 Nanomedicine and nanoparticles: general overview

The concept of nanotechnology is rapidly becoming integrated with everyday life. The notion of "nano" (one-billionth of a meter) is no longer considered science fiction, but represents an emerging technology with broad application not only within the biological sciences. In the past few years this multidisciplinary field of nanotechnology has undergone revolutionary development. Nanoscience has a huge potential to bring benefits in areas as diverse as drug development, water decontamination, information and communication technologies, and the production of stronger, lighter materials. Human healthcare nanotechnology research can result in immense health benefits thanks to advances across medicine, communications, genomics, and robotics. In particular, the term nanomedicine refers to highly specific medical interventions at a molecular scale for curing diseases and repairing damaged tissues. The application of nanotechnology to healthcare holds great promise for revolutionizing medical treatments, imaging, faster diagnosis, drug delivery, and tissue regeneration.[4,5] A NP is defined as the smallest unit (10^{-9} meters) that can still behave as a whole entity in terms of properties and transport and is able to cross biological membranes and access cells, tissues, and organs that larger-sized particles normally cannot.[6] Nanoparticle systems in cancer therapies provide better penetration of therapeutic and diagnostic agents and a reduced risk in comparison to conventional treatments.

Interestingly, research in the field of molecular oncology has shown important characteristics of tumor cells, referred to as the "hallmarks of cancer", which can aid in finding more efficacious therapies. Tumor angiogenesis is one of the most studied areas. Several studies have shown that tumor vasculature possesses structural anomalies that impede drug delivery. Among these, the most relevant hallmark is their poor architecture with an abnormal basement membrane and fissures between the endothelial cells due to an absent pericyte lining. This state of leaky vasculature accompanied by a poor lymphatic drainage system causes a differential interstitial pressure at the center of tumors compared with that at the periphery. Due to this pressure difference, molecules ranging from approximately 10 to 100 nm preferentially accumulate in the tumor and are retained for longer, unlike the uncoated drugs, which are much smaller and are cleared by the kidneys. This phenomenon is called the "enhanced permeability and retention effect" (EPR). The retention time of drugs packed in NPs is ten times higher than that of unpacked drugs, which eventually return to the vascular system.[7] Hence, this EPR effect attributed to the leaky vasculature is considered a boon for drug-delivery systems within the nanosize range.

The majority of studies in this field are performed in experimental settings, both in vitro and in animal models, but their translation into the clinics is difficult due to human variability and safety-related issues. Drug delivery nanosystems have been produced with the principal aim of protecting therapeutic agents and improving their biodistribution and therapeutic index. Based on the chemical nature of the preparations, nanomaterial-based agents that are specific for nanomedicine include mainly polymer or lipid-based carriers such as liposomes and NPs as nanospheres, nanocapsules, micelles, dendrimers, nanocrystals, nanogolds, carbon NPs, silica NPs, ceramic NPs, chitosan NPs, low-density lipoproteins (LDL), nanoemulsions, and quantum dots. Drugs can be absorbed onto the surface, entrapped inside, or dissolved within the matrix of these vehicles. The first application of a targeted nanosystem for drug delivery concerned liposomes and was reported in 1980.[8] Actual researche has progressed to the development of multifunctional NPs able to respond to the environment, so facilitating a more effective drug delivery. Also, the shape helps functionality. The diversity of delivery systems allows NPs to be developed with a diverse array of shapes, sizes, and components, which enables them to be tailored for specific applications. However, the primary consideration when designing any drug-delivery system is to achieve more effective therapies, by controlling the drug concentration in the therapeutic window, reducing cytotoxic effects, and improving patient compliance.

Currently, NPs can be incorporated into cells and this technology has been evaluated to manipulate and track stem cells.[9] Interactions between nanocarriers and cells have already been studied. Nanocarriers can be combined into cells using transfection agents, which are mostly cationic, positively charged molecules such as poly-L-lysine or polyethyleneimine.[10,11] However, these transfection agents are toxic and have not been clinically approved. More recently, spontaneous incorporation of NPs into cells has been described without the use of any transfection agents.[12–19] Futhermore, NPs have been shown to enter inside cells via passive transport[20] and active endocytosis.[14,21,22] Three types of endocytosis pathways can be used according to the size of the NP: clathrin-mediated endocytosis, caveolae-mediated endocytosis, and clathrin- and caveolae-independent endocytosis.[23] Once inside the cells, NPs are usually transported to the endolysosomal system, where they are destroyed. Polylactide and polyglycolide NPs as well as lipid nanocapsules (LNCs) are able to escape the lysosomal compartment by disrupting the integrity of the lysosome membrane. In the acidic pH of secondary endosomes, the surface charge of polyglycolide or polylactide NPs changes from anionic to cationic resulting in a local NP–membrane interaction and escape of NPs into the cytoplasm.[16,17] For LNCs, the hypothesis is that the hydroxystearate polyethyleneglycol (PEG), which constitutes the most external phase of LNCs, has a lysosomotropic property and therefore destabilizes the lysosome membrane.[24] This endolysosomal escape leads to NP accumulation in the cytoplasm.

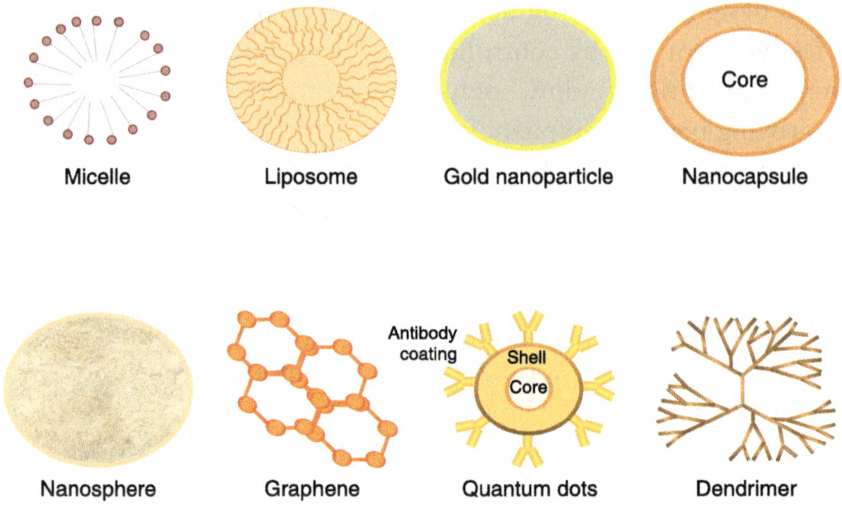

Figure 4.1 Schematic representation of the main nanotechnology-based drug-delivery systems

The most extensively studied NPs are polymer NPs, micelle NPs, liposomes, gold NPs, silver NPs, metal oxide, magnetic NPs, carbon nanotubes, fullerenes, peptides, silica NPs, graphene NPs, quantum dots, and dendrimers (Figure 4.1).

4.2 Main types of nanoparticles for drug-delivery systems

4.2.1 Polymer nanoparticles

Polymer NPs were developed to create delivery systems with excellent drug and protein loading and release properties, a long half-life, and little toxicity. They are widely used for the encapsulation of a large number of bioactive molecules and drugs, so they are differently synthesized according to the needs of the application and the type of drug to be encapsulated. Polymer NPs are formulated using emulsion/solvent evaporation or solvent displacement techniques.[25] Using these methods, a variety of therapeutic agents including both low-molecular-weight lipophilic or hydrophilic drugs and high-molecular-weight DNA or antisense oligonucleotides (AONs) can be encapsulated in polymer NPs.[26,27] These NPs can be structured as nanospheres or nanocapsules. Nanospheres are characterized by a matrix system in which the drug is dispersed, whereas the nanocapsules have a reservoir in which the drug is confined in a hydrophobic core surrounded by a single polymeric membrane (core–shell structure). These carriers show a high stability in biological fluids and resist the enzymatic metabolism, effectively permeating through cell membranes. The core matrix of these NPs is formulated using various biodegradable polymers and copolymers, such as the synthetic poly(lactic-coglycolic acid), poly(alkylcyanoacrylate), and poly(ε-caprolactone) (PCL), or the natural chitosan, albumin, gelatin, alginate, and collagen. The degradation drug release rate of these polymers can be controlled by adjusting their molecular mass from days to months, and in the case of copolymers, this is possible by modifying their composition and microstructure.[28] Polymer NPs have been used as transport vectors to deliver various peptides into the central nervous system (CNS) by intravenous injection, such as hexapeptide dalargin, loperamide, tubocurarine and doxorubicin. Poly-D,L-lactide-coglycolide (PLGA) is one of the most investigated and successfully used biodegradable nanosystems because it undergoes hydrolysis in the body to produce its biodegradable metabolite monomers: lactic and glycolic acids. Surface modification of PLGA, drug encapsulation methods, particle size, so as additives added during formulation, drug molecular weight, and the ratio lactide: glycolide ratio have a strong influence on the release and effective response of formulated nanomedicines. For drugs of acidic nature, PLGA monomers are combined with alginate, chitosan, pectin, poly(propylenefumarate) polyvinylacohol, and poly(orthoester). The antineoplastic drug paclitaxel promotes the polymerization of tubulin causing cell death by disrupting the cell division process. It is one of the potent anticancer agents, used as first choice for chemotherapy against primary ovarian carcinoma as well as breast and colon cancers, but possesses poor solubility. PLGA combined with vitamin E and tocopheryl PEG succinate has been used to encapsulate paclitaxel. This formulation has shown better results in comparison with the traditional formulation. Using some additive with the PLGA-NPs, 100% drug encapsulation efficiency was achieved with full antitumor activity.[29] Another efficacious antitumor drug is cisplatin, but it is toxic for healthy tissues. An in vitro preparation composed of cisplatin and PLGA–methoxy(polyethylene glycol) (mPEG) prepared by double emulsion methods revealed prolonged drug half-life in the bloodstream following intravenous administration.[30] Tamoxifen prevents proliferation of precancerous cells, competitively binding to tumor estrogen receptors to produce a nuclear complex that reduces DNA synthesis and inhibits estrogen effects. Tamoxifen loaded polyethylene oxide modified PCL was prepared by a solvent displacement method. About 90% drug encapsulation efficiency has been achieved loading tamoxifen in the ratio of 10% by weight of polymer. polyethylene oxide–PCL NPs exhibited a significantly increased level of accumulation of the drug within the tumor as well as extended presence in the systemic circulation.[31] The PEG–PCL amphiphilic block copolymeric nanospheres containing taxol have an outer shell of mPEG and a hydrophobic inner core of PCL. They are reported to show promising anticancer activity and could be potentially useful as a novel delivery system for the anticancer drug taxol.[32] Polymers can also be used to coat other types of NPs. Polyethylene glycol is a hydrophilic polymer that, when used to coat the surface of NPs, allows them to avoid clearance by the reticulo-endothelial system and to cross the BBB thanks to receptor-mediated phagocytosis and passive leakage through permeable capillaries in tumors.[33] Other hydrophilic polymers including hydrogel (polyacrylamide), dextran, and polysorbate have been used with NPs to extend the plasmatic half-life of drugs and improve their delivery across the blood–brain barrier (BBB).[33]

The polymer–drug conjugation is formed between side-chain grafting of drugs to polymer chains, allowing them to deliver high doses of chemotherapeutic drugs. These agents bear numerous functional groups that are available for covalent binding to a variety of biochemically active groups, which allow them to have different simultaneous functions acting on several tumor targets.[34] Nanoconjugates that carry more than one functional group provide the capability to inhibit several tumor pathways, deliver optimal drug concentrations to the site of treatment, and reduce adverse effects on healthy tissue at the same time. Nanoconjugate polymers are generally synthesized around a polymer with pendant functional groups like –OH, –COOH, or –NH_2. As an example, prolindac (AP5346) is composed of a hydroxypropyl methacrylate backbone copolymer with platinum grafted to the side chains through a pH-sensitive chelator designed for drug release in the tumor environment. Preclinical data demonstrate superior efficacy of the polymer–drug conjugates, using multiple cancer models including an M5076 sarcoma platinum-resistant tumor xenograft mouse model, multiple colon xenograft models, L1210 leukemia, and 0157 hybridoma models.[35] Polyamino acids grafted with drugs on the side chains are another class of polymer–drug conjugates with great efficacy.[36] In the case of polyglutamate–

glycine–campthotecin (CT-2106), degradable linkers have allowed drug loadings ranging from 5% to 50%.[37] Nanoconjugates can overcome the drawbacks of conventional chemotherapy such as drug resistance and toxicity by specifically targeting tumor cells, activating cancer cell uptake, and bypassing multidrug-resistance transporters. Xyotax, a similar polymer–drug conjugate (polyglutamate–paclitaxel), is used in several clinical trials including prostate cancer, metastatic breast cancer, neck cancer, and metastatic colorectal cancer. The clinical data show an improvement in median survival in patients given Xyotax compared with the control group. One benefit of the treatment was the reduction of multiple side effects including neurotoxicity.[38]

4.2.2 Micelle nanoparticles

Micelle nanoparticles (MNPs) are amphiphilic spherical structures composed of a hydrophobic core and a hydrophilic shell. The hydrophobic part is the inner core of the block copolymer which encapsulates the drug, whereas the outer hydrophilic shell protects the drug from the aqueous environment and stabilizes the MNPs against recognition in vivo by the reticulo-endothelial system. The core can sometimes be made up of a water-soluble polymer that is rendered hydrophobic by the chemical conjugation of a water-insoluble drug, and by complexation of the two oppositely charged polyions. The polymer always contains a nonionic water-soluble segment and an ionic segment that can be neutralized by an oppositely charged surfactant to form a hydrophobic core. The electrostatic interaction between the ionic segment of the block polymer and the surfactant group changes these segments from water-soluble to water-insoluble, leading to a hydrophobic core in the micelles.[39] MNPs can be engineered by means of ligand coupling, or addition of pH-sensitive moieties, according to the biological characteristics of their target.[40] Once the target site is reached, micelles are internalized into the cells via fluid-state endocytosis. To overcome permeability problems, amphiphilic copolymers are used to encapsulate poorly water-soluble anticancer drugs in MNPs. These have an inner core made up of hydrophobic block copolymer in which the drug becomes entrapped, and an outer shell of hydrophilic block copolymer that reduces the interactions of drugs with the outer aqueous environment, keeping them stable. The hydrophilic outer part can be made up of polyethers like PEG, and poly(ethylene oxide). Other hydrophilic shells are made up of polymers such as poly(acryloylmorpholine), poly(trimethylene carbonate), and poly(vinylpyrrolidone). Genexol-MNP is the first non-targeted polymeric micellar formulation of paclitaxel and cisplatin approved for cancer therapy. It is currently being evaluated in a clinical phase II trial in the USA for the treatment of advanced non-small-cell lung cancer. The clinical phase II results showed ~ 30% of the patients had a stable disease status and 60% of the patients had an increased survival of 1 year.[41]

Many recent studies have revealed that polymer-conjugated drugs and NPs show prolonged circulation in the blood followed by passive accumulation in tumors, even in the absence of targeting ligands, demonstrating the existence of a passive retention mechanism. The characteristic tumor vasculature features – leaky endothelial cells, increased vascular tortuosity and aberrant basement membrane – render tumor blood vessels permeable to macromolecules. Hence, numerous studies have shown that the EPR effect causes passive accumulation of macromolecules and NPs in solid tumor, enhancing the therapeutic index while decreasing side effects. The delivery of drugs can be made more target-specific by using biological interactions such as antigen–antibody binding or locally applied signals such as sonication or heating. Active targeting takes advantage of overexpression of cell surface tumor-associated antigens as well as of tumor-specific antigens and the relatively more acidic nature of tumor compared with normal tissue. Active targeting decreases adverse side effects because the drug accumulates only in the tumor sites, and it allows cellular uptake of the drug through endocytosis.

Surfactants are being incorporated into anticancer metal-based drugs. The surfactant dodecylamine reacts with selenious acid to produce a quaternary ammonium salt, which can be conjugated to copper or cobalt ions to form copper or cobalt cationic complexes. Initial studies demonstrated effectiveness in vitro against five human monolayer tumor cell lines namely MCF7 (breast carcinoma), HEPG(2) (liver carcinoma), U-251 (glioma), HCT116 (colon carcinoma), and H-460 (lung carcinoma). Recent evaluation has been undertaken of the potential antitumor activity of NK012, a 7-ethyl-10-hydroxycamptothecin (SN-38) micellar formulation, and bevacizumab in human lung cancers.[42] Nude mice bearing PC-14 or A549 lung adenocarcinoma xenografts show evidence of significant tumor growth inhibition compared with saline controls. B-Lapachone (b-lap) is a novel anticancer agent, whose cell-killing effect is activated by the enzyme NADPH-quinone oxidoreductase 1 (NQO1), a flavoprotein that is overexpressed in breast, prostate, and lung cancers.[43] In cancer cells where NQO1 is overexpressed, the agent undergoes futile cycling, resulting in the generation of reactive oxygen species (ROS). Experimental studies have demonstrated that growth inhibition occurs in cells overexpressing NQO1, whereas cells in which NQO1 is absent are unaffected at equivalent concentrations. Antitumor efficacy was examined in female nude mice bearing subcutaneous A549 lung tumors and orthotopic Lewis lung carcinoma. Following intravenous administration of b-lap micelles, A549 tumor growth suppression was shown. In the Lewis lung carcinoma model a doubling of survival was observed (16 days compared with 8 days in controls).[44] Another target for molecular cancer therapy is heat-shock protein 90 (HSP 90), a molecular chaperone, which under normal conditions is responsible for prevention of protein aggregation.[45] HSP 90 becomes overexpressed under conditions of stress, resulting in tumorigenesis and increased proliferation in a variety of cancers including lung, prostate, and breast. Tanespimycin, a derivative of the HSP 90 inhibitor geldanamycin, has been explored clinically for chemotherapeutic purposes. The mechanism of action of tanespimycin involves the degradation of oncogenic signaling proteins, inducing cell death via apoptosis. In patients with multiple myeloma, treated with tanespimycin, disease stabilization was observed.[45]

4.2.3 Liposomes

Liposomes are vesicles made up of a lipid bilayer, resembling a cell membrane. The lipids form a bilayer based on hydrophobic interactions in continuous parallel packing, with the hydrophilic head groups positioned towards the aqueous environment. They possess advantages of carrying hydrophilic, lipophilic, as well as amphoteric drug molecules, either entrapped inside or on their micellar surface. The brain distribution of long circulating liposomes can be modulated by conjugation of appropriate targeting vectors. Examples of brain targeting vectors include monoclonal antibody (to anti-transferrin receptor, or to insulin receptor), cationized proteins (cationized human serum albumin), endogenous peptides or plasma proteins. The basic mechanism by which these liposomes achieve brain concentration by crossing the BBB is by coupling with brain drug transport vector through absorptive-mediated transcytosis, or by receptor-mediated transcytosis. Hence, by manipulating the liposome structures, they can be constructed to be temperature or pH sensitive to permit controlled release of their contents. The dual problems of mediating BBB transport and inhibiting peripheral clearance of liposomes were solved by the combined use of PEGylation technology and chimeric peptide technology.[46] After surface modification of liposomes with these substances, they behave as if sterically stabilized, due to enhanced hydrophilicity imparted by the polymers' hydrophilic chains, a lower contact angle between particles and phagocytic cells, and due to the lesser interaction between serum opsonins, thereby preventing opsonization. Constructed temperature-sensitive liposomes loaded with doxorubicin in combination with local hyperthermia, show a complete regression of human tumor xenografts in all the mice studied.[47] The encapsulation of doxorubicin in PEG-coated liposomes (Doxil/Caelyx, or PLD), was developed to enhance the safety and efficacy of conventional doxorubicin. The liposomes alter pharmacologic and pharmacokinetic parameters of conventional doxorubicin, so that drug delivery to the tumor is enhanced while the toxicity normally associated with conventional doxorubicin is decreased. In preclinical models, PLD produced remission and cure against many cancers, including tumors of the breast, lung, ovaries, prostate, colon, bladder, and pancreas, as well as lymphoma, sarcoma, and myeloma. PLD appeared to overcome multidrug resistance, possibly as the result of increased intracellular concentrations and an interaction between the liposome and P-glycoprotein function.[48] Several phase II studies showed promising activity of PLD in recurrent ovarian cancer patients with response rate ranging from 16 to 25%.[49]

Bevacizumab is a recombinant humanized monoclonal antibody that inhibits vascular endothelial growth factor (VEGF), a growth factor ligand responsible for angiogenesis.

Results from several phase III clinical trials comprising colorectal, non-small cell lung and breast cancer, demonstrate that bevacizumab results in superior patient response rates. Bevacizumab can be used as a targeting moiety to enhance the NPs efficacy. For this reason, bevacizumab-labeled cationic liposomes have been developed, to improve targeting to several pancreatic cancer cell lines including Capanel, HPAFeII, and PANCe1.[50] Bevacizumab-conjugated liposomes had modest impacts on cell viability in vitro, and demonstrated increased cellular uptake by PANCe1 cells grown in the presence of VEGF. Protein stabilization of liposomes is being studied to deliver hydrophobic drugs such as docetaxel for cancer therapy. Docetaxel is encapsulated into the liposome bilayer and stabilized by albumin to prevent rapid drug leakage (ATI-1123). The results of ATI-1123 efficacy studies in human xenograft mouse models for prostate, pancreatic, and non-small-cell lung cancer showed partial tumor regression in 90% of the PC3 tumor xenograft model, and improved efficacy in the pancreas model.[51] Small-interfering RNA (siRNA) fragments have been found to suppress gene expression, with immense silencing efficiency and relatively low toxicity. The siRNAs degrade extremely rapidly in physiological environments and are eliminated almost immediately from circulation upon injection. Liposomes prove ideal carriers for biological agents such as siRNA because of their stable aqueous core. Moreover, it is possible to combine RNA-interfering strategies with traditional chemotherapeutics. One example is the Raf/mitogen-activated protein kinase kinase (MEK)/extracellular signal-related kinase (ERK) pathway, which is essential for cellular proliferation, and found to be aberrant in several cancers.[52] As a result, several inhibitors of key proteins in the cascade have been developed as potential chemotherapeutics. Recently, it has been demonstrated that liposomes encapsulating a

Mcl1-specific siRNA (siMcl1) and a chemical MEK inhibitor (PD0325901) showed a valid antitumor efficacy in vitro and in vivo. Following encapsulation and complexation of PD0325901 and siMcl1 respectively, the liposomal formulation was administered to KB cells. Western blot results showed that co-delivery of both agents significantly reduced expression of Mcl1 and pERK1/2 proteins.[53] Antisense therapy represents a gene silencing strategy that stands to make a profound impact on cancer therapy. In a phase I study, a liposomal formulation, LErafAON, that encapsulates the raf antisense oligonucleotide, was administered with the purpose of acting on c-raf, a protein that bestows cancer cells with resistance to radiation or chemotherapy. In patients with advanced solid tumors undergoing radiation therapy, the c-raf-1 messenger RNA was inhibited in three, four exhibited partial response, four had stable disease, and four showed progressive disease.[54] Recently, the use of bisphosphonates, such as zoledronic acid, was explored as a treatment strategy, given its ability to inhibit the release of growth factors essential for cancer cell growth and differentiation in bone. Emerging data from several clinical trials serve to highlight a potential anticancer effect of zoledronic acid, as well as chemotherapeutic synergy with established drugs.[55] However, zoledronic acid has an extremely rapid blood clearance and preferential accumulation in bone, necessitating encapsulation in NPs. Lipo-ZOL is a liposomal formulation of zoledronic acid that increases circulation times, reduces accumulation in bone, and increases targeting to tumors.[56]

4.2.4 Gold nanoparticles

Gold nanoparticles (GNPs) exhibit unique physicochemical properties, including the ability to bind amine and thiol groups, allowing surface modification and use in biomedical applications. GNPs are used to prepare nanoshells composed of gold and copper, or gold and silver, to function as contrast agents in magnetic resonaonce imaging (MRI), and gold–silica for photothermal ablation of tumor cells. Classically, GNPs enter into cells with a nonspecific receptor-mediated endocytosis mechanism.[57] In vivo GNPs passively accumulate at tumor sites that have leaky immature vasculature with wider fenestrations than normal mature blood vessels. Difficulties in using the EPR effect for tumor drug delivery exist owing to the heterogeneity of tumor vasculature, particularly at the centre of poorly differentiated cancers, as well as particle detection and uptake by the reticulo-endothelial system. PEGylation represents the most common method of reducing reticulo-endothelial system uptake, producing a hydrated barrier causing steric hindrance to the attachment of phagocytes. GNPs have also been used for cancer cell imaging and targeting. In various clinical trials the 27-nm citrate-coated GNPs bound with thiolated PEG and tumor necrosis factor-α (CYT-6091) (Aurimmune; CytImmune Sciences, Rockville, MD, USA) has shown an increase of tumor targeting.[58] An important feature of GNPs is their capacity to absorb and scatter specific wavelengths of light across the visible and near-infrared (NIR) spectra. The most useful nanoshells have a silica core diameter of around 120 nm, with a 10-nm layer of gold shell, and they absorb NIR light (800 nm) and can create intense heat that is lethal to cells. An in vivo study demonstrated that 100-nm gold nanoshells maximally accumulated in SK-BR-3 human breast tumors 24 hours after intravenous injection. When a laser tuned to the nanoshell resonance was applied, average tumor temperatures increased by 9 °C in control mice, and 37 °C in nanoshell-treated mice, with irreversible tissue damage in the nanoshell group. All mice in the nanoshell group survived 90 days with no evidence of tumor recurrence.[59] Positive results in vivo were also obtained with photothermal ablation therapy in a mouse model for colon carcinoma after intravenous administration of PEG-coated gold nanoshells.[60] Gum arabic-stabilized GNPs are used for diagnostic and therapeutic applications, showing optimal in vitro and in vivo stability. The compound is nontoxic, distributes minimally to nontarget organs in biodistribution studies, and produces contrast on computed tomography imaging.[61] A study group has shown an approach for imaging and targeting cancer cells using dendrimer entrapped GNPs. These dendrimer entrapped GNPs, which when covalently linked to folic acid and fluorescein isothiocyanate molecules are stable, hydrophilic, biocompatible, and able to specifically bind to cancer cells that overexpress high-affinity folate receptors. The folic acid-conjugated NPs are subsequently endocytosed into lysosomes of cancer cells, providing a means for targeting and imaging of these cells.[62] An interesting new therapeutic strategy foresees the connection of antibodies–nanoshells is able to target cancer cells by interacting with specific surface antigen expressed only by tumor cells. The benefit of the nanoshell-mediated approach is that the energy can pass through the healthy tissue and leave the neighboring cells intact while killing only the tumor cells that have been targeted by the nanoshells.

4.2.5 Silver nanoparticles

Silver nanoparticles are part of the emerging nanotechnology that have gained increasing interest in the field of nanomedicine because of their particular properties and therapeutic potential in treating a large variety of diseases.[63,64] The biological activity of silver has been attributed to the presence of the silver ion, Ag^+. Silver NPs inhibit the VEGF-induced angiogenesis in retinal endothelial cells. Alteration of the permeability barrier integrity plays a major role in drug-based therapies, as well as in the pathogenesis of cardiovascular diseases, inflammation, acute lung injury syndromes, and carcinogenesis. Recently the molecular mechanism of silver NPs on VEGF-induced and interleukin-1β (IL-1β) -induced retinal endothelial cell permeability has been evaluated. Both VEGF and IL-1β increase endothelial cell permeability via an Src-dependent pathway. Silver NPs were found to block VEGF- and IL-1β-induced permeability in retinal endothelial cells from porcine retina, and this inhibitory effect was dependent on the modulation via Src phosphorylation at Y419.[65] A novel study has demonstrated the antitumor activity of biologically synthesized silver NPs in a Dalton's lymphoma ascites tumor system in vitro, by activation of the caspase 3 enzyme, which is known to have a potent inhibitory effect on disease progression in a mouse model, leading to a potent restorative effect in the treated tumor volume.[66]

4.2.6 Metal oxide nanoparticles

Titanium dioxide (TiO_2) is a semiconductor, well-known as an ultraviolet light (UV) -inducible catalyst in the photo-oxidation of organic substrates and the deactivation of bacteria, algae and viruses.[67,68] Under UV excitation, TiO_2 NPs of various sizes and morphologies have been reported to exhibit cytotoxicity toward some tumors.[69,70] One recent example[69] describes 50-nm rhodamine-labeled TiO_2/PEG constructs that can be internalized into rat glioma C6 cells. The antitumor performance was evaluated in glioma cell spheroids representing a provisional three-dimensional model valuable for translation to animal xenografted models. The cytotoxic effect of the UV-irradiated photocatalyst depended on the concentration of TiO_2/PEG and the light exposure time. More than 90% of cells were killed by a UV dose of 13.5 J/cm^2 in the presence of the nanocatalyst at a concentration of 0.5 mg/mL. Moreover, fluorescent images of the photocatalyst-treated spheroids co-stained with apoptosis and necrosis markers, Annexin V-fluorescein isothiocyanate and propidium iodide, reveal the prevalence of induced apoptotic cell death within the first 6 hours. Functionalization of 5-nm high-crystallinity TiO_2 NPs with a monoclonal antibody recognizing IL-13 receptor (IL-13R) fostered NP delivery specifically to glioblastoma (GBM) cells in a manner dependent upon cellular membrane IL-13R expression. The direct visualization of the TiO_2–antibody/receptor interaction and mapping of the IL-13R location and distribution throughout a single A172 brain cancer cell was demonstrated using synchrotron-based X-ray fluorescence microscopy.[71,72] It is well established that UV-photoexcitation of bare TiO_2 particles in aqueous solution results in the formation of various ROS, mainly hydroxyl (-OH), peroxy (-HO_2) radicals, and singlet oxygen (-$1O_2$)[73] However, in the case of dopamine- and dopamine-antibody-modified TiO_2 particles, ROS arise from multiple, mechanically distinct redox chemistries, and the principal ROS produced is the superoxide anion, formed by reaction of photogenerated electrons with molecular oxygen.[74] Furthermore, in cell studies of photo-induced cytotoxicity toward A172 glioma cells in the presence of selective ROS, quenchers were consistent with these results.[71] Nanostructured porous TiO_2 has been developed as a biocompatible nano-device for constant chemotherapy drug release into the CNS.[73] A porous titanium carrier uploaded with low concentrations of a cytostatic platinum complex was capable of inducing DNA fragmentation, possibly via a strong interaction between nitrogen atoms in nucleotides, and Lewis acid sites on both the titanium surface and the platinum complex coordination sphere. Application of this material directly on to C6 glioma xenografted into Wistar rats resulted in a significant decrease in tumor size and growth rate.

4.2.7 Magnetic nanoparticles

One of the most frequently used, non-invasive imaging tools for disease diagnosis and monitoring is MRI, including in cancer. Imaging techniques that can selectively image proliferating cells in vivo can provide critically important insights into tumor growth rate, degree of tumor angiogenesis, effectiveness of treatment, and vigor of normal cells. Contrast agents that are commonly used in clinical practice for the brain and spinal cord MRI are based on gadolinium. However, a major problem associated with MRI is its low sensitivity. Utilization of nanotechnology to improve the sensitivity and efficacy of MRI for cancer detection and imaging is an area that researchers have focused on in the last several decades. Magnetic NPs, used in biomedical applications mainly, have an inorganic NP core and in most cases are coated by a suitable coating material. Suitable coatings not only increase the stability and solubility of the nanoformulation, but can also be used to incorporate a targeting moiety to increase the imaging sensitivity and to do real-time monitoring. Enhanced proton relaxation is one of the most added-value properties that make magnetic NPs one of the best contrast agents for biomedical applications of MRI. Iron oxide and superparamagnetic iron oxide (SPIO) -NPs exhibit magnetic properties, which are used for MRI and also provide an opportunity to control particle transport by external magnets. Superparamagnetic iron oxide contrast agents either form the core of magnetic NPs that have a polymeric coating, or are more homogeneously integrated into polymeric NPs.[75] The signal intensity of these NPs is related to the size of the particle, its position, its concentration within a given voxel, data acquisition parameters, the magnetic field, and the dosage of the SPIO-NP.[76] SPIO-NP has been used as a bowel contrast agent (Lumerin,

Gastromark) and for spleen/liver imaging (Endorem, Feridex). Macrophage-specific uptake of SPIO-NPs increases the contrast between healthy and diseased tissue because most liver tumors are devoid of it. Negative enhancement effects of SPIO-NPs on T1/T2-weighted MRI sequences allowed increased lesion conspicuousness and increased lesion detection compared with non-enhanced imaging. It is well documented that with the help of this technique, liver tumors or metastases as small as 2–3 mm can be detected. Through conjugation of iron oxide NPs with hydrophilic polymer coatings, such as dextran or PEG, it is possible to obtain a steric configuration that prevents opsonization of NPs in the serum and a reduction of their uptake by the reticulo-endothelial system.[77] Recently, antibiofouling polymer-coated magnetic NPs as nanoprobes for MRI have been characterized. SPION were coated with the protein-resistant or cell-resistant polymer, 3(trimethoxysilyl) propyl methacrylate-τ-poly(ethylene glycol) methyl ether methacrylate), to generate stable, protein-resistant MRI probes. The compound could detect tumors in vivo using MRI, and can be used as a potentially efficient cancer diagnostic probe.[78] MNPs exhibit acute toxicity in vivo, which has limited their clinical translation. Oxidative stress and interference with mitochondrial energy production by MNPs can lead to cytotoxicity.

4.2.8 Carbon nanotubes

Carbon nanotubes (CNs) are essentially cylindrical molecules made of carbon atoms. They are synthesized by rolling sheets of graphene into hollow tubes that are single-walled (SWNTs) (0.4–2 nm in diameter), double-walled (1–3.5 nm in diameter), or multi-walled (MWNTs) (2–100 nm in diameter). CNs can be synthesized by heating carbon black and graphite in a controlled flame environment. One of the main advantages of the CN is its ability to deliver drugs directly to cancer cells. It has also been suggested that CNs could be used as nanocarriers for delivering drugs into the body via injectable routes.[79] Drugs can either attach to the outer surface of the CNs via functional groups, or be loaded inside the CNs. Attachment of the anticancer drug to the outer surface of the CNs can be through either covalent or noncovalent binding, including hydrophobic, π–π stacking, and electrostatic interactions.[80] The mechanism by which CNs enter cells is unclear. The evaluated processes are the passive diffusion of CNs through the lipid bilayers of the cell membrane, and the attachment of CNs to the external cell membrane, resulting in its absorption by the cell, using an energy-dependent process. Generally speaking, small CNs with a length of up to 400 nm are internalized by a diffusion mechanism, whereas CNs of 400 nm in length are internalized by endocytosis.[81] Functionalization and alteration of CNs and other graphite nanoplatfom surface chemistry can reduce or eliminate complement activation, while making the CNs more biocompatiable.[82] Functionalized SWNTs were conjugated with paclitaxel through branched PEG chains via a cleavable ester bond. The resultant formulation was more effective in suppressing tumor growth in vivo than Taxol or paclitaxel-PEG conjugated in a 4T1 breast cancer animal model.[83] Similar findings have been obtained when paclitaxel was loaded into PEGylated SWNTs or MWNTs using HeLa cells and MCF-7 cancer cells lines.[84] Kam et al.[85] have shown the possibility to direct nanotubes to specifically targeted cancer cells by using coating of the nanotube surface with folic acid. In this way CNs bind specifically to cancer cells that over-express folate receptors, and then allow receptor-mediated endocytosis of nanotubes. With this approach, it is possible to introduce genes directly into tumor cells without any cellular or viral vector by using aerosol, systemic delivery or microcellular injection. Another interesting use of CNs, is characterized by their ability to carry siRNA molecules that exert RNA interference on target gene expression.[86] The authors used siRNA-conjugated CNs that specifically targeted murine telomerase reverse transcriptase, and show that delivery of siRNA into tumor cells silences the target gene, inhibits the proliferation of cancer cells in vitro, and suppresses tumor growth.[86] CNs are also able to absorb light in the NIR region resulting in heating of the nanotubes.[87] Engineering the structure of MWNTs, by creating intentional surface defects or dopants, will cause scattering in the travelling current and also increase the heating of the nanotube. This physical feature of the engineered MWNTs can be employed to thermally destroy the tumor cells by using MWNTs that have good heat-conducting properties. Although the toxicity of CNs is not fully understood and toxicity study results are conflicting, it is important to be aware of potential complications. It has been noted that as the particle size decreases, the surface area of the particles increases. This means that there will be more area available for chemical interactions to take place, which would enhance the toxicity of the particles. A novel research report shows that when murine epidermal cells were exposed to unpurified SWNTs containing 30% iron, significant dose-dependent activation of transcription factor activating protein-1 occurred.[88] Systemic application of CNs can result in oxidative stress in end organs, and inhalational exposure to CNs can result in acute lung injury, inflammation, and fibrosis.[89]

4.2.9 Fullerenes

Fullerenes are a family of carbon allotropic compounds in the form of a hollow sphere, ellipsoid or tube. The most common form is C60. It has also led to the discovery or synthesis of other fullerene variations, such as C70, C20 (the smallest member), CNs (elongated, tube-structured fullerene), carbon nano-onions, and nano buds.[90] An important property of the C60 molecule is its high symmetry. Fullerenes have the ability to assume different forms and to encage compounds. The unique physical, chemical, electrical, and optical properties of fullerenes and their derivatives have led to their incorporation into new or improved devices and materials, and to advancements in engineering, industry, and science. However, the difficult processibility of fullerenes has presented a major problem in the hectic search for medicinal applications. C60 is insoluble in aqueous media and aggregates very easily. Commonly, fullerenes are encapsulated in special carriers like cyclodextrins, calixarenes, polyvinylpyrrolidone, micelles and liposomes. A second technique is that of chemical functionalization with amino acid, carboxylic acid, polyhydroxyl group, and amphiphilic polymers to increase the hydrophilicity. Fullerenes and their derivatives show potential antiviral activity. The antiviral activity of fullerene derivatives is based on several biological properties, including their molecular architecture and antioxidant activity. Another potential medical application of fullerenes is related to their photo-excitation. In fact, fullerene can be excited from ground state to 1C60 by photo-irradiation. In the presence of molecular oxygen, the fullerene can decay from its triplet to ground state, transferring its energy to O_2, generating a single oxygen $1O_2$, and is highly cytotoxic. Again, in the presence of oxygen, the fullerene radical anion can transfer one electron, producing a superoxide anion radical $O_2{\bullet}^-$ and a hydroxyl radical •OH.[91] Iwamoto and Yamakoshi[92] introduced a highly water-soluble C60-N vinylpyrrolidine copolymer as agent for photodynamic therapy. Liu et al.[93] demonstrated the use of a PEG-conjugated fullerene containing Gd^{3+} ions for photodynamic therapy (PDT) in combination with MRI. The authors demonstrate, through experimental data, that tumor PDT effect was significantly promoted by photosensitizer tumor targetability and MRI activity. By attaching hydrophilic moieties, fullerenes become water-soluble and are capable of carrying drugs and genes for cellular delivery. A lipophilic slow-release drug delivery system that employs fullerene derivatives to enhance therapeutic efficacy in tissue culture was designed by Zakharian et al.[94] So modified fullerenes have the potential to provide a significant anticancer activity in cell culture, as demonstrated with a C60–paclitaxel conjugate. Research continues into ways to increase the solubility of fullerenes and to investigate the toxicity of fullerenes and their derived compounds.

4.2.10 Peptides

Peptides that specifically interact with receptors overexpressed by cancer cells have been successfully developed as targeting molecules for drug delivery and in vivo imaging.[95] The interaction of peptides and proteins with the cell membrane results in their penetration into the cell, or the formation of pores within the cell membrane. Because of their ability to target and enter cells, peptide and protein carriers hold great potential for the delivery of genes and AONs to cancer cells.[95] Bombesin peptide and its analogs can be used to target gastrin-releasing peptide (GRP) receptors. Hence, in vivo GRP receptors are overexpressed in GBM, small cell lung, gastric, pancreatic, prostate, breast, cervical, and colon cancers. Recently, GNPs functionalized by a high load of thioctic acid–bombesin peptide were used to target prostate tumor xenografts in SCID mice. Using normal and prostate tumor-bearing mice, they showed that this compound exhibits a high binding affinity to the tumor, and confirmed that these constructs are GRP receptor specific and accumulate with high selectivity in GRP-receptor-rich pancreatic acine.[96] Fibroblast growth factor analogs can be used to target cells expressing fibroblast growth factor receptors (FGFRs). This receptor family is often expressed both on tumor cells and in neovasculature. Truncated human basic fibroblast growth factor peptide (tbFGF) was recently used to achieve targeting of liposomes carrying chemotherapeutic drugs.[97] This peptide contains both the bFGF receptor binding site and a part of the heparin-binding site, which allows it to bind FGFRs on a cell surface without stimulating cellular proliferation. Somatostatin and its analogs can be used to target somatostatin receptors that are overexpressed in both small-cell and non-small-cell lung cancers. Albumin-bound paclitaxel (Abraxane) is currently being tested as a first-line therapy, or in combination with other drugs, for metastatic breast cancer and other cancers that have been shown to be sensitive to taxane drugs, such as ovarian and prostate cancers. Preclinical studies have shown that the concentration of paclitaxel, bound to albumin in endothelial cells and in the extravascular space, was significantly increased. Peptides can also act as therapeutic agents conjugated to NPs. Melittin is a cytolytic peptide that represents a potential candidate for cancer chemotherapy. Melittin is a 26-amino-acid α-helical peptide derived from the venom of the honeybee *Apis mellifera*. It is a nonspecific cytolytic peptide that attacks all lipid membranes, leading to significant toxicity when injected intravenously. The basis of melittin's action is a physical and chemical disruption of membrane structure resulting in a profound compromise of the cell permeability. Recently, it has been demonstrated that synthetic nanoscale vehicles like perfluorocarbon NPs can deliver melittin by flexible passive and active molecular targeting to kill both established solid tumors and precancerous lesions. This study has proposed a novel linking strategy to generate biocompatible peptide–nanostructures for lipidic nanocarriers, including perfluorocarbon nanoemulsions, liposomes, as well as cells for combined molecular imaging and cell-targeted therapeutics.[98]

4.2.11 Silica nanoparticles

Silica is major component of sand and glass, and it has been used in the synthesis of NPs. Functional groups can also be added to the surface, making them appealing for designs for different applications. Immunofluorescent labeling of both a cancer cell surface marker and tissue sections by dye-doped silica NPs has demonstrated high specificity and high intensity.[99] Different strategies have been explored for using silica NP probes to target cancer cells. Affinity and specificity associated with the antibody–antigen recognition have been studied in developing immunoassays. Primary or secondary antibodies are covalently immobilized onto the NP surface to selectively and efficiently bind various cancer cells.[100] In one study, a mouse anti-human CD10 antibody was used as the recognition element on NPs. Fluorescence microscopy was then used to image the leukemia cells. The brightly fluorescent cells bound with NPs were easily detected under the fluorescence microscope.[101] Other affinity reagents, such as receptor ligands and recognition peptides, can also be attached onto NPs to label cell-membrane proteins. For instance, folic acid was attached to dye-doped silica NPs and targeted to SCC-9 cancer cells, which overexpress folate receptors.[102]

Peptide-targeted uptake is another efficient technique for cancer cell imaging. This technique is based on the propensity of the cells to recognize and internalize NPs labeled with specific peptides, and even deliver them to specific cellular compartments. Human lung adenocarcinoma (A549) cells (in vitro) and rat brain tissue (in vivo) were successfully labeled using transactivator of transcription-labeled NPs. Using this strategy, diagnostic and therapeutic agents can be delivered to the biological target of interest.[102] Recently, aptamers have emerged as a novel class of ligands. Aptamers are short strands of DNA/RNA for recognition of a variety of targets, including proteins and small molecules, as well as complex samples. Aptamers have significant advantages over antibodies and peptides, including high affinity, excellent specificity, and lack of immunogenicity. Specific targeting of acute leukemia cells with aptamer-conjugated NPs has been developed using fluorescence microscopy or flow cytometry.[103] NP–aptamer conjugates greatly increase the fluorescence signal from the cell. This property shows the potential applications of silica NPs in the elucidation of cells with low densities of aptamer binding sites, or with relatively weak binding probes where the fluorescence signal from the fluorophore is too weak for observation.[104] Tris(2,2′-bipyridyl)dichlororuthenium(II) hexahydrate (RuBpy)-doped silica NPs have been used as highly sensitive and photostable labels in Affymetrix GeneChips technology. Biotin-labeled complementary RNA samples from a human lung cancer cell line were hybridized on the arrays, and then incubated with streptavidin and staining with PEG-biotin-labeled NPs. Even with the present unfavorable imaging modality and existing optical excitation and detection systems of the GeneChips, the fluorescent silica NPs were demonstrated to be superior to the traditional streptavidin–phycoerythrin. Fluorescent silica NPs can act as nonviral vectors for gene delivery and biophotonics methods, and may be used to optically monitor intracellular trafficking and gene transfection. The potential of cationic silica NPs was investigated for in vivo gene transfer.[105] The NPs were tested for their ability to transfer genes in vivo in the mouse lung, and a twofold increase in the expression levels was found with silica particles in comparison to enhanced green fluorescent protein alone. Silica NPs are also promising candidates for improved drug delivery systems because of their intrinsic hydrophilicity, biocompatibility, and nontoxicity, as well as the excellent protection they provide for their encapsulated drugs. With drug molecules loaded into silica NPs, surface modification of the NPs with bio-recognition entities can allow specific cells or receptors in the body to be located. Upon target recognition, NPs can then release their drug payload at a precisely controlled rate by tailoring the internal structure of the particles according to a desired diffusion (release) profile. The high surface area (>900 m^2/g), tunable pore diameter, and uniform mesoporous structure of the mesoporous silica NPs offer unique advantages for loading and releasing large quantities of biomedical agents. Mesopores loaded with guest molecules were capped by inorganic NPs, or large organic molecules, via a chemically cleavable disulfide linkage to the mesoporous NP surface. Since drug molecules are effectively physically trapped, they are unable to leach out of the mesoporous NP host therefore preventing any premature release. Compared with many current biodegradable polymer-based drug delivery systems, that rely on the hydrolysis-induced erosion of the carrier structure, the mesoporous NP structure provides the ability to release the cargo in a controlled manner.[106,107]

4.2.12 Graphene nanoparticles

Graphene is an innovative two-dimensional nanomaterial possessing a particular chemical configuration, and unique physical, electronic, optical, thermal, and mechanical characteristics.[108–112] It is a carbon allotrope with a bidimensional hexagonal structure and, together with its related derivatives, such as its oxide (GO), has shown potentials not only in the field of nanoelectronics,[113] composite materials,[114–118] energy technology,[119–121] sensors,[122] and catalysis,[123–128] but has also found promising applications in biomedical research. The importance of the functionalization of GO has been the key for the biological and biomedical applications of graphene.[124,129–131] Graphene oxide is a versatile material for various applications, which range from targeting controlled drug/gene delivery, photothermal and photodynamic cancer therapy, and biological sensing and imaging, to multifunctional nanoplatforms.[132] Even in the field of nanotechnology, nano-GO (NGO) has shown enormous potential derived from its simple chemical structure and morphology. The intensive research on the biomedical applications of graphene and its derivatives is due to its intrinsic biocompatibility, low cost of production, and ease of the processes of functionalization.[127,128] However, as experimental material, tests of its toxicity in vitro and in vivo have produced controversial results. The biomedical applications of graphene represent a new field with significant potential. In oncology, NGO has been studied for the treatment of different tumor types, both as native molecule, and as drug delivery vehicle.[129–136] As an example, non-targeted NGO was tested for the treatment of primary tumors given its enhanced permeability and retention effect.[137] Moreover, tailoring its surface with specific ligands as targets for proteins expressed in cancer cells, nontargeted NGO could improve its role in tumor treatment. Instead, other research was conducted in targeted nanomaterials to detect, visualize, and destroy cancer cells with minimal side effects on normal cells.[138] Nano-GO has also been experimentally employed as a bioimaging agent in different techniques, including MRI and radionuclide-based imaging.[139–141] Graphene nanoplatelets, namely single layers of carbon atoms having nanoscale lateral dimensions, are insoluble and possess difficult interfacial interaction with the targeting matrix.[142] They absorb insoluble molecules via noncovalent binding,[143] with low drug-loading efficiency. The coating of silica NPs, good vectors for insoluble chemotherapeutic drugs, on a graphene nanosheet could improve the interfacial properties of graphene.[144,145] It has also been considered that the morphology of different glioma cells, the molecular status of cell lines, and the concentration of graphene may regulate the biological response of tumoral cells. One of the first attempts at using graphene for the treatment of gliomas employed its property as carrier. The chemotherapeutic drug 1,3-bis(2-chloroethyl)-1-nitrosourea (BCNU), widely used for the treatment of brain tumors, has poor thermal stability and a short half-life. Immobilization of BCNU on a nanocarrier might increase its stability and extend its half-life. Nano-GO was conjugated with polyacrylic acid to improve the aqueous solubility and increase the cell penetration efficacy, and was used as nanocarrier for BCNU. This nanocarrier significantly prolonged the half-life of bound BCNU from 19 to 43 hours and showed efficient intracellular uptake by GL261 cancer cells. The in vitro anticancer efficacy of polyacrylic acid–GO–BCNU was demonstrated by a 30% increase in DNA interstrand cross-linking and a 77% decrease in the half maximal inhibitory concentration of the drug.[146] The photothermal activity of graphene has also been investigated in the treatment of brain tumors. The mechanisms of graphene-mediated photothermal killing of cancer cells apparently involved oxidative stress and mitochondrial membrane depolarization, resulting in mixed apoptotic and necrotic cell death characterized by caspase activation/DNA fragmentation and cell membrane damage, respectively. Despite lower NIR-absorbing capacity, a suspension of polyvinylpyrrolidone-coated graphene sheets exposed to NIR radiation generated more heat than CNs under the same conditions. Subsequently, graphene NPs performed significantly better in inducing photothermal death of U251 human glioma cells in vitro. The superior photothermal sensitivity of graphene sheets could be largely explained by their better dispersivity.[144] A more recent study combined the chemo-photothermal targeted therapy of glioma within one novel multifunctional drug delivery system using a targeting peptide (IP) -modified mesoporous silica-coated graphene nanosheet (GSPI). Doxorubicin was conjugated with the GSPI-based system (GSPID), showing synergistic chemo-photothermal properties. Cytotoxicity experiments demonstrated a higher rate of death of glioma cells compared with that of single chemotherapy or photothermal therapy. Furthermore, the IP modification could significantly enhance the accumulation of GSPID within glioma cells.[144] A study by Jaworski et al.[145] examined the toxicity of graphene platelets in U87 and U118 glioma cell lines. They revealed that graphene is toxic to glioma cells, indicating the potential applicability of graphene platelets in glioma therapy. Because of their large surface area, graphene platelets did not enter the glioma cells, but adhered to them. The mechanism the authors explained concerns the different biological impacts of graphene electrons on various cell and tissue types, and the subsequent interaction with cell membranes and receptors, blocking the supply of nutrients, inducing stress, and activating apoptosis. This direct interaction between platelets and tumor cells, caused by their chemical and physical properties, indicates that this effect might depend on the surface morphology of the cells.[145] In fact, the properties of glioma cells may vary depending on the cell line.[146] Exposure to graphene induced apoptosis in both glioma cell lines but with different results: 68% in U87 and 99% in U118 cells. The activation of apoptosis was also observed in a study on the influence of graphene and nanotubes on the U251 glioma cell line, which was treated photothermally. Moreover, mechanisms of necrosis were also present – 24% in U87 cells and 0.2% in U118 cells. The U118 cells have reduced expression of the anti-apoptotic gene *Bcl2* and of the suppressor gene *PTEN*, allowing for apoptosis, in contrast with U87 cells. The precise mechanism of apoptosis is still unknown. However, it is clear that because of a mutation of the gene *p53* in U118 cells, this gene is not involved in apoptosis activation.[147]

4.2.13 Quantum dots

Quantum dots (QDs) are structurally colloidal semiconductor nanocrystals, ranging from 2 to 10 nm in diameter. QDs can be synthesized from various types of semiconductor materials via colloidal synthesis or electrochemistry. The most commonly used QDs are cadmium selenide (CdSe), cadmium telluride (CdTe), indium phosphide (InP), and indium arsenide (InAs). These NPs have unique photophysical properties, such that upon excitation they emit fluorescence that is brighter and more stable than that of traditional fluorophores, and their size can be varied to achieve excitation and emission at different wavelengths. QDs can be used as probes for high-resolution molecular imaging of cellular components, for tracking cell activities and movement inside the body, for specific targeting interaction through antibodies linked onto the NPs surface, with specific tumor-associated antigens expressed on the cancer cell surface. This interaction permits penetration inside targeted cancer cells of specific drugs, protein, siRNA, genetic materials, and AONs and modulates genic expression into the cancer cell genome. In vivo cancer targeting and imaging in living animals by QDs was first demonstrated by Gao et al.,[148] wherein both subcutaneous injection of QD-tagged cancer cells (prostate cancer) and systemic injection of multifunctional QD probes were used to achieve sensitive and multicolor fluorescence imaging of cancer cells. The utility of the QD–aptamer–doxorubicin conjugateas a novel targeted cancer imaging, therapy, and sensing system has been demonstrated in a recent study. The targeted QD imaging system (QD-Aptamer) was capable of differential uptake and imaging of prostate cancer cells that express the prostate-specific membrane antigen. The Tan et al.[149] study group conjugated siRNA targeting the gene encoding human epidermal growth factor receptor-2 (hEGFR-2) to QDs by using these last ones not only as carriers, but also as a means to monitor the transfection efficiency. By directing antibodies against EGFR-2 overexpressed by breast cancer cells, it was possible to induce a selective interaction of siRNA–QD conjugates with cancer cells, and a receptor-mediated endocytosis of conjugates and subsequent silencing effects on the target gene through RNA interference. CdSe particles may leak cytotoxic cadmium ions after long-term exposure to UV light, whereas CdTe particles produce ROS as a result of the loss of their protective coating after long-term circulation.

4.2.14 Dendrimers

A dendrimer is generally defined as a macromolecule that is characterized by its highly branched three-dimensional structure, which provides a high degree of surface functionality and versatility.[150] The generation number and chemical composition of the core, branches, and surface functional groups, determine the size, shape, and reactivity of dendrimers. Dendrimers have attracted attention as possible drug carriers because of their unique properties, namely their well-defined three-dimensional structure, the availability of many functional surface groups, their low polydispersity, and their ability to mimic. Dendrimers can function as drug carriers either by encapsulating drugs within the dendritic structure, or by interacting with drugs at their terminal functional groups via electrostatic or covalent bonds (prodrug).[151] Dendrimers have been studied extensively for targeting and delivery of therapeutic agents for cancer and of contrast agents for MRI. The avidimers are dendrimers targeted to tumor vasculature using a methotrexatepolyamidoamine (PAMAM) bioconjugate platform functionalized with small targeting ligands.[152] The authors demonstrated in vitro, that drug-free dendrimer conjugates were not cytotoxic, and that drug-loaded dendrimer conjugates had no effect on folate receptor-negative cells. Polyanionic PAMAM dendrimers showed rapid serosal transfer rates in crossing an adult rat intestine in vitro, and had low tissue deposition. The transport of PAMAM and surface-modified PAMAM across a cell monolayer follows endocytosis-mediated cellular internalization. However, non-biodegradable dendrimers may potentially accumulate in lysosomes depending on their frequency and dose of administration. Various studies report that PEG-modified dendrimers show reduction of cytotoxicity and immunogenicity, high exocytosis rate and low accumulation in endothelial cells, with excellent solubility and favorable pharmacokinetics.[153] Amine-terminated polyamidoamine (PAMAM) dendrimers appear to be an ideal class of building blocks for developing multifunctional gene vectors. Angiopep is a high brain penetration peptide that targets to the LDL receptor-related protein-1. Ke et al.[154] coupled angiopep to

PEGylated PAMAM dendrimer G5.0 via the distal end of PEG, and used it to deliver pEGFP-N2 plasmid to the brain both in vitro and in vivo. The plasmid DNA covalently labeled with fluorescent dye, ethidium monoazide bromide, was detected in the brain of the mice treated with the PAMAM-PEG-Angiopep/DNA. Multimodal dendrimer-conjugated magnetofluorescent nanoworms, called dendriworms, were developed recently for siRNA delivery.[155] The magnetic core in dendriworms enables in vivo imaging of dendriworms with MR whereas PAMAM dendrimers conjugated to the magnetic core allow nucleic acid delivery and targeting. Dendriworms accumulate in the lungs and the reticuloendothelial filtration organs following systemic delivery. Dendriworms administered with convection-enhanced drug delivery, efficiently delivered EGFR siRNA to suppress the expression of EGFR in GBM tumors in a mouse model.[155] Yang et al.[156] prepared EGF-carrying boronated PAMAM dendrimer G4.0 for neutron capture therapy of brain tumors. Doxorubicin was conjugated to arginyl-glycyl-aspartic acid-coupled PEGylated PAMAM dendrimer, via a degradable disulfide spacer, for controlled release in the treatment of glioma tumors.[157] PAMAM dendrimers have been tested as genetic material carriers. SuperFect-DNA complexes, a transfection reagent consisting of activated dendrimers, are characterized by high stability and provide more efficient transport of DNA into the nucleus. The high transfection efficiency of dendrimers may be due to their well-defined shape but also to the low pK of the amines.[158]

4.3 Application in brain tumors

Nanomedicine devices are ideal for delivering drugs to brain tumors, loading them as carriers via a variety of chemical methods including encapsulation, adsorption, and covalent linkage. In brain tumor treatment, various molecules at different steps and pathways, such as cell immortalization and apoptosis escape, tumor neoangiogenesis, and invasion of normal tissues, have been studied as possible targets of a novel therapeutic model. Recent advances in molecular, biological and genetic diagnostic techniques have evidenced new cerebral glioma-associated biomarkers and their implications for glioma progression. The possibility to block the more contemporary pathway into glioma by molecular-based targeted approaches, using a nanocarrier loaded with anti-cancer agent, represents an interesting therapeutic strategy to overcoming the BBB and delivering drugs and/or genetic probes into brain tumor cells in a selective manner. The future challenges of this approach may be the possibility to modify the cell genome and induce it to a reversion into the wild-type conditions, the enhancing of immune system antitumor capacity, and the targeted drug delivery into brain tumor cells.

4.3.1 Emerging studies

Bernardi et al.[159] evaluated the efficacy of immunonanoshells in vitro against medulloblastoma and malignant glioma cell lines. They used an antibody against the human epidermal growth factor receptor (HER-2) to target gold–silica nanoshells to medulloblastoma cells, demonstrating cell death in the HER-2-overexpressing medulloblastoma cell lines, after exposure to laser light. In glioma cell lines, they showed the capacity of these immunonanoshells in causing leading cell death in U373 and U87 malignant glioma cell lines. This mechanism may also be innovative in diagnostics. In fact, iron-oxide NP-based MRI contrast agents, targeted NP-based MRI contrast agents, and intraoperative

NP-enabled brain tumor delineation, may produce contrast enhancement at an earlier stage of gliomas.[160] Another emerging specific contrast agent is represented by a nanoprobe that targets gliomas that may express membrane-bound matrix metalloproteinase-2. This nanoprobe, named chlorotoxin-conjugated superparamagnetic nanoprobe (PEG-coated NPs), has the capacity to selectively detect neoplastic cells in gliomas, medulloblastoma, prostate cancer, sarcoma, and intestinal cancer.[161] Wang et al.[162] in a recent study have demonstrated, using a molecular targeting of glioma cells through CD133 antigen overexpressed on the surface of GBM cells, a prominent photothermal selective damage of targeted glioma cells. Molecular targeting in this case has been performed using CNs, conjugated with anti-CD133 monoclonal antibodies. The efficiency of liposomal and other nanoplatform systems has been enhanced by various molecular targeting, such as IL-13 expression, transferring receptor and LDL surface receptor pathways. Glioma cells show upregulation of expression of IL-13Rα2 on their surfaces. In a recent study, the improvement of internalization of doxorubicin-loaded nanoliposomes, targeted with conjugated IL-13, and cytotoxicity in U251 glioma cells has been shown. In an in vivo animal model the authors demonstrated the inhibition of the growth of subcutaneously implanted gliomas.[163] In anticancer gene therapy, the efficiency of liposomes has been increased through surface ligand targeting, via monoclonal antibodies to specific receptors that are upregulated on glioma cell surfaces, such as transferring receptors, LDL receptors, IL-13R. A biopolymeric gene delivery NP has recently been shown to be effective in vivo in delaying tumor growth. This polymeric NP-based nonviral gene delivery vector is a cationic albumin-conjugated pegylated NP, in which is incorporated a plasmid encoding proapoptotic Apo2 ligand/

tumor necrosis factor-related apoptosis-inducing ligand (Apo2L/TRAIL). After intravenous injection of plasmid-loaded NPs, and subsequent accumulation in a C6 murine glioma model, incorporation of plasmid DNA into the host cell genome and inhibition of tumor growth lead to prolonged survival in mice bearing implanted C6 gliomas.[164] An alternative therapeutic strategy is the PDT with targeted delivery systems. PDT involves the intratumoral release of photosensitizers combined with local excitation by an appropriate wavelength of light, resulting in the production of oxygen and other ROS, which initiate apoptosis and cytotoxicity and microvascular injury, within treated neoplastic tissues. The recent molecularly targeting approach leads to the selective detection of cancer cells, through molecular recognition processes, such as ligand–receptor or antibody–antigen interaction. PDT is an interesting approach for the treatment of cerebral gliomas, resulting in a very selective loco-regional therapeutic approach with an important improvement in local control of tumors and a significantly improved survival.[165] A recent in vitro and in vivo study has shown the efficacy of indomethacin-loaded nanocapsules in significantly reducing the tumor size of implanted glioma in rats. Moreover pathological analysis demonstrated the lack of some important malignant characteristics typical of GBM, such as mitotic index and microvascular proliferation.[166]

Gold nanostructures demonstrate great potential in imaging in diagnostics serving as molecular contrast agents. Because of their high extinction coefficient gold NPs can be used as contrast agents for dark-field, light-scattering, and two-photon luminescence imaging. Moreover, gold nanomaterials were used for signal amplification in photoacoustic tomography (PAT). PAT is an advanced diagnostic hybrid technique based on laser-induced thermoelastic expansion through biological tissue, which allows the combination of the benefits of optical and ultrasound imaging.[167] PAT allowed noninvasive in vivo molecular imaging of the vascular system in the brain of a live small animal and, even more, imaging of important hallmarks of tumor development and progression such as angiogenesis[168] and hypoxia.[169] Wang and coworkers[167] applied PEGylated gold nanoshells as exogenous NIR contrast agent for laser-induced PAT of the rat brain in vivo with high spatial resolution and satisfactory sensitivity.

Nanotechnology nonviral gene delivery systems, such as CNs, represent an interesting therapeutic choice. These systems carry siRNA molecules that exert RNA interference on target gene expression after their internalization into the target tumor cells and release. Zhang et al.[86] with in vitro studies, used siRNA that specifically targeted murine telomerase reverse transcriptase and showed that delivery of siRNA via CNs into tumor cells, silenced the target gene, inhibited the proliferation of cancer cells in vitro, and suppressed tumor growth in mouse models. Higher therapeutic index per NP system application in brain tumor treatment will need better engineering for higher loading and better controlled release of the drug into a tumor site, and an improvement in the development of most functional drugs. A new interesting and promising strategy to achieve localized drug delivery to tumor tissue is magnetizing targeting. This approach has the advantage that the accumulation and retention of drug-loaded magnetic NPs in cancer cells can be enhanced, by the attraction of NPs to the tumor location using an externally applied magnetic field.[170] Schneider et al.[171] have examined a "double-punched" approach to overcome the escape of GBM cells to immune surveillance, through an active specific immunization with Newcastle disease virus-infected tumor cells, and blocked the transforming growth factor-β (TGF-β) production by delivery of TGF-β AONs using polybutyl cyanoacrylate NPs. This approach induced a significant decrease in plasma TGF-β_2 level, as well as an increase in the rate of high-affinity IL-2R (CD25) on lymphocytes and consequently of antitumoral cytotoxicity. By using anti-focal adhesion kinase phosphotothioate AONs packaging into liposomes in U251 microglia cells, the downregulation of expression levels of FAK and the activation of apoptosis, through increase in caspase-3 activity, a key mediator of apoptosis in mammalian cells, has been shown.[172] Paclitaxel, one of the most successful anticancer drugs, is the first of a new class of microtubule stabilizing agents and has demonstrable antitumor activity in glioma cell lines. However, because of the poor aqueous solubility and low therapeutic index of paclitaxel, its clinical application is extremely limited. Furthermore, it is reported that the activity of paclitaxel against brain tumors has been disappointing in a phase II study, because of drug resistance and poor penetration across the BBB.[173] A recent study demonstrated that drug-loaded mPEGylated PCL long-circulating NPs provided a sustained release of the embedded drug, and higher, or at least comparable, in vitro cytotoxicity to that of Taxol injection against C6 GBM cells.[174] It has been demonstrated that an Angiopep-2-modified drug delivery system could enhance delivery of a gene drug and a NIR fluorescent probe across the BBB.[175] The potential therapeutic effect of Angiopep-conjugated PEGePCL NPs, loaded with paclitaxel as a dual-targeting drug delivery system in the treatment of glioma, was evaluated in a recent study. PEGePCL NPs were conjugated to Angiopep for enhanced delivery across the BBB as well as for targeting the tumor via LDL receptor-related protein-mediated endocytosis. Angiopep-conjugated PEGePCL NPs were internalized by U87 microglia (U87MG) glioma cells, and displayed higher cell uptake and stronger inhibition and apoptosis toward glioma cells due to LDL receptor-related protein-mediated endocytosis. As well as this, the Angiopep-conjugated PEGePCL NPs construct increases the transport of the NPs across the BBB, and target the brain glioma by the in vitro coculture model and in vivo imaging of brain fluorescence.[175] In an ongoing phase I clinical trial, paclitaxel albumin-stabilized NP formulation is being used in treating advanced cancers such as bladder cancer, brain and CNS tumors. The authors demonstrated, in a subset of patients, a decrease in tumor vascular permeability.[176] Majoros et al.[177] evaluated a multifunctional dendrimer conjugated with fluorescein isothiocyanate (for imaging studies), folic acid (for targeting cancer cells overexpressing folate receptors), and paclitaxel (chemotherapeutic drug). The authors demonstrated in vitro that drug-free dendrimer conjugates were not cytotoxic, and that drug-loaded dendrimer conjugates had no effect on folate receptor-negative cells. It is known that NPs may trigger an inflammatory process, resulting in the release of different pro-inflammatory cytokines, chemokines, ROS, and transcription factors that could explain some known side effects. These mechanisms may involve microglia and interact in different ways with normal brain tissue and glioma cells. All of these aspects are crucial in modulating an effective strategy for a molecular therapeutic approach, trying to hit an important molecular pathway, such as neovascularization, invasiveness and interaction between tumor and perilesional tissue. The more recent experimental studies on the molecular approach in glioma treatment have been performed by hypothesizing a single molecular target. This strategy has shown poor results, as demonstrated by the actual pharmacological and molecular multimodal strategy of treatment. The authors think, that using the intrinsic capacity of NPs should be very interesting in attempting to structure a new NP-based molecular approach against two or more molecular targets, contemporaneously.[178–184]

In a laboratory approach, the authors are trying to create an engineered carrier, loaded with an antisense molecule against hypoxic ischemic factor-1α (HIF-1α) and IL-8. The HIF-1α appears to be highly involved in the development of a characteristic tumor phenotype influencing growth rate, invasiveness, and metastasis. HIF-1α actively regulates downstream processes, and is also itself influenced by the tumor microenviroment in many different ways. As a result, local hypoxia, due to increased proliferation or insufficient oxygen supply, inactivation of tumor suppressors, oncogenes, and growth factors, along with other cell types, such as macrophages, contributes to form a tumor microenviroment that is capable of modulating the HIF response itself. Antisense inhibition of HIF may be a strong target for antiangiogenic therapy. The authors study group has recently shown high expression levels of prostaglandin E1 synthase and IL-8 in high-grade glioma cells and microglial cells, strongly correlate with the grading of a tumor.[187,188] During progression of gliomagenesis, leukocyte infiltration and necrosis are two biological phenomena associated with the development of neovascularization. In malignant gliomas, IL-8 further localizes in oxygen-deprived cells surrounding necrosis. Macrophages are known to produce high levels of IL-8, which has a tumorigenic activity, by inducing tumor growth and angiogenesis. Tumor pseudopalisading cells secrete HIF, which induces IL-8 secretion. The IL-8-binding chemokine receptors CXCR1, CXCR2 and the Duffy antigen receptor for chemokines (DARC) were found in all astrocytoma grades by reverse transcription–polymerase chain reaction analysis. These results support a model in which IL-8 expression, by induction of inflammatory stimuli, may be an early step in astrocytoma development. It seems that augmented IL-8 directly and/or indirectly promotes angiogenesis by binding to DARC, and induces leukocyte infiltration and activation by binding to CXCR1 and CXCR2. The contemporary actions of IL-8 into glioma angiogenesis and leukocyte infiltration, as well as macrophages, microglial cells, and ECM components involvement, suggest IL-8 as a future interesting target in brain tumor treatment.

Nanoparticle systems, can represent ideal devices for the delivery of specific compounds to brain tumors across the BBB.[189–192] Nanotechnology and nanomedicine have been used to form new therapeutic intracerebral drug delivery systems, and to develop treatments for various diseases and disorders. By using nanotechnology in drug design and delivery, it will be possible to deliver the drug to the targeted tissue and cells across the BBB, to release the drug at a controlled rate, and to be able to escape from degradation processes. Solid tumors require therapies to actively penetrate deeply into the tumor in order to affect a large proportion of cancer cells. Nanotechnology provides a unique advantage in glioma therapy because the size scale is of the order of the proteins used for cell function. The size and shape of NPs can be tuned to exert a desired therapeutic response on a specific target.

4.3.2 Antiangiogenic approach

Antiangiogenic approaches have been extensively exploited to provide a rationally designed therapy for the treatment of malignant gliomas. The brain tumor endothelium, with characteristics of high proliferation, high permeability, and high expression of proangiogenic factors, is a particularly appealing therapeutic target for this strategy.[185,186]

Antiangiogenic approaches in glioma therapy have been strongly directed against a VEGF pathway. In an in vivo murine model, created by implantation of U87MG malignant glioma cells in mice, Im et al.[187] demonstrated the suppression of the ability of glioma cells to form tumors in mice. This result was obtained after transfection of antisense VEGF complementary DNA, in an antisense orientation through the recombinant adenoviral vector Ad5CMV-αVEGF. Infection of U87MG malignant glioma cells resulted in the reduction of the level of the endogenous VEGF messenger RNA, and in reduced production of the VEGF targeted secretory form. Agemy et al.[188] have proposed a multifunctional theranostic NP in which the CGKRK peptide provides the targeting function that takes the NPs to tumor vascular cells and into their mitochondria. The NP uses the mitochondria-targeted D[KLAKLAK]$_2$ peptide as the drug and iron oxide, as a diagnostic component for MRI. In addition, the NP was combined with the tumor-penetrating peptide i arginyl-glycyl-aspartic acid, which enhances the NP penetration into the extravascular tumor tissue. Systemic treatment of GBM-bearing mice with this compound eradicated most tumors in one GBM mouse model, and significantly delayed tumor development in another. An important molecular target used to selectively detect glioma cells is IL-13, based on upregulated expression of IL-13α2 on the surface of GBM cells. In a recent study, Madhankumar et al.[163] showed the improvement of internalization of doxorubicin-loaded nanoliposomes targeted with conjugated IL-13, compared with nontargeted nanoliposomes in U251 glioma cells. In an in vivo animal model, the authors demonstrated growth inhibition of subcutaneously implanted gliomas. Bernardi et al.[159] evaluated the efficacy of immunonanoshells in vitro against both medulloblastoma and malignant glioma cell lines. In this study, the authors, using gold–silica nanoshells coated with an antibody against HER-2 to target medulloblastoma cells, showed cell death in the HER-2-overexpressing medulloblastoma cell lines after exposure to laser light. The same authors conjugated gold–silica nanoshells to an antibody specific to IL-13Rα2, which is strongly expressed in gliomas, demonstrating that these immunonanoshells are capable of leading to cell death in U373 and U87 malignant glioma cell lines. Convection-enhanced delivery (CED) techniques were developed to address the diffusion-limited penetration of agents directly delivered to the brain. This strategy has been used to deliver proteins and small particles, including liposomes and polymeric NPs, into the brain. CED provides penetration through a large volume of brain tissue but it is limited by unpredictable drug distribution and potentially high intracranial pressures. Combining polymeric controlled release with CED could improve the drug distribution limitations of implantable wafers while also offering spatiotemporal distribution control that is lacking from CED. Poly(lactic-co-glycolic acid) is capable of encapsulating and releasing a variety of agents, including chemotherapy drugs, for long periods of time. The authors evaluated the efficacy of CED of surface-modified, drug-loaded PLGA NPs to treat intracranial glioma using the topoisomerase I inhibitor camptothecin. The camptothecin is an attractive drug for delivery by controlled release because it has known anticancer activity, but is limited by low solubility and serious systemic toxicity. The NPs were shown to be effective both in culture and in vivo, with a statistically significant survival benefit observed in all animals treated.[189] CED of iron oxide NPs in a mouse glioma model results in MRI contrast of the NPs and effective intratumoral and peritumoral distribution of NPs in the brain. A significant therapeutic effect was found after CED of both iron oxide NPs and EGFRvIIIAb-iron oxide NPs in mice. Dispersal of the NPs over days, after the infusion has finished, may potentially target infiltrating tumor cells outside the tumor mass that are potentially responsible for tumor recurrence and the demise of patients. Use of bioconjugated magnetic NPs may permit the advancement of CED in the treatment of malignant gliomas because of their sensitive imaging qualities on standard T2-weighted MRI and therapeutic effects.[190] Cyclo-oxygenase (COX)-2 is the key enzyme in arachidonic acid metabolism resulting in prostaglandin production, and is induced by several factors, such as growth factors, cytokines, and tumor promoters. In particular, COX-2 expression and prostaglandin production are associated with tumorigenesis and tumor progression. Celecoxib, a selective COX-2 inhibitor, has been reported to mediate growth inhibitory effects and to induce apoptosis in various cancer cell lines. PLGA NPs incorporating celecoxib were prepared for antitumor drug delivery. The PLGA NPs incorporating celecoxib showed the same cytotoxicity against U87MG tumor cells as celecoxib itself. Furthermore, celecoxib did not affect the degree of migration of U87MG cells. When C6 rat glioma cells were used, PLGA NPs incorporating celecoxib showed dose-dependent cytotoxicity similar to that of celecoxib itself. Neither celecoxib nor PLGA NPs incorporating celecoxib affected COX-2 expression in C6 cells on a Western blot assay.[191] Curcumin is a polyphenolic compound derived from the Indian spice turmeric. NanoCurc™, a recently described polymeric NP formulation of curcumin, was used to treat medulloblastoma and GBM cells. This formulation caused a dose-dependent decrease in growth of multiple brain tumor cell cultures, including the embryonal tumor-derived lines DAOY and D283Med, and the GBM neurosphere lines HSR-GBM1 and JHH-GBM14. The reductions in viable cell mass observed were associated with a combination of G2/M arrest and apoptotic induction. Curcumin also significantly decreased anchorage-independent clonogenic growth and reduced the CD133-positive stem-cell-like population. Levels of signal transducer and activator of transcription 3 were also attenuated. These data suggest that curcumin NPs can inhibit malignant brain tumor growth through the modulation of cell proliferation, survival and stem cell phenotype.[192]

4.3.3 Gene therapy

Gene therapy has the potential to effectively treat cancer by treating the root of the disease. This technology involves the delivery of DNA molecules to cancer cells to insert or modify a gene in an effort to treat the disease. The delivery of DNA can be accomplished using a variety of vectors including viruses, cell-based systems, and synthetic vectors. For glioma gene therapy, viral vectors have been used to deliver suicide genes, pro-apoptotic genes, p53, cytokines, and caspases. These studies have shown promising preclinical results, but clinical trials have been limited by the fact that transduced cells were found only within a very short distance of the delivery site. To overcome these limitations, synthetic vectors have been developed to more safely deliver DNA. In this study the authors investigate targeted gene delivery to C6 glioma cells in a xenograft mouse model using chlorotoxin-labeled NPs. The developed nanovector consists of an iron oxide NP core, coated with a copolymer of chitosan, PEG and polyethylenimine. The chlorotoxin promotes specific uptake of nanovectors into glioma cells, exposing a higher proportion of target cells to the delivered payload. These results could provide insight into the design of more effective gene delivery vehicles for improved treatment outcome of gene therapy for glioma.[161] In brain tumor treatment, the efficiency of liposomes as nonviral gene delivery vectors has been increased through surface ligand targeting, via monoclonal antibodies specific to certain receptors upregulated on glioma cell surfaces, such as transferrin receptors, LDL receptors, and IL-13 receptors. A biopolymeric gene delivery NP has recently been shown to be effective in vivo in delaying tumor growth. This polymeric NP-based nonviral gene delivery vector is a cationic albumin-conjugated pegylated NP that incorporates a plasmid encoding proapoptotic Apo2 ligand/tumor necrosis factor-related apoptosis-inducing ligand (Apo2L/TRAIL). After intravenous injection of plasmid-loaded NPs in a C6 murine glioma model, plasmid DNA is incorporated into the host cell genome, so inhibiting tumor growth, with prolonged survival.[164] Another gene therapy approach is the injection into cancer cells with genes that can destroy the cells. A pro-drug or an inactive form of a toxic drug is administered to the patients, and this drug will kill off any cancer cells with the suicide genes in them. The use of specific NPs can represent a valid alternative to overcome possible toxic and infective effects of classic gene therapy. The use of NP nonviral gene delivery systems such as CNs blocks a selective genic function without toxic effect on cellular phenotype. These systems have the ability to carry short single genes, and also short DNA fragments or siRNA molecules that exert RNA interference on target gene expression after their internalization.[193] One of the most promising new strategies is the RNA interference-based approach, wherein small double-stranded RNA molecules can sequence specifically to inhibit the expression of targeted oncogenes. To harness the full potential of this approach, the prime requirements are to deliver the siRNA molecules with high selectivity and efficiency into tumor cells, and to monitor both siRNA delivery and the resulting knock-down effects at the single cell level. Herein, the authors describe the synthesis and target-specific delivery of multifunctional siRNA-QD constructs for selectively inhibiting the expression of EGFR variant III in target human U87 GBM cells, and subsequently monitoring the resulting downregulated signaling pathway with high efficiency. This study demonstrates the multifunctional siRNA-QD strategy focusing on targeted delivery, high transfection efficiency, and multimodal imaging/tracking. These novel methods and applications complement recent advances in nanomaterial-based siRNA delivery, nanomaterial-based molecular imaging, and siRNA-based chemotherapeutic strategies reported recently. This strategy could also provide highly useful information regarding biosurface chemistry of nanomaterials. In addition, the application of multifunctional siRNA-QDs to modulate the key cancer signaling pathways is important not only for a selective chemotherapeutic strategy but also for dissecting signaling cascades triggered by inhibiting specific proteins.[193]

4.3.4 Other potential approaches

Although immunotherapy is being investigated as an adjunct treatment, the ability of gliomas to escape the immune response will continue to be a significant obstacle to this strategy. One approach to overcome the local immunosuppressive tumor microenvironment is the activation of the innate immune system by Toll-like receptor agonists such as CpG oligonucleotides (CpG). Because the Toll-like receptor 9, CpG receptor, is located intracellularly, the authors have hypothesized that methods that enhance CpG internalization may also potentiate its immunostimulatory response. In this study, it has been reported that CNs enhanced CpG uptake by tumor-associated phagocytic cells, and resulted in their activation both in vitro and in vivo. Furthermore, a single injection of low-dose CN–CpG complexes eradicated intracranial gliomas through activation of natural killer and CD8 cells. These findings demonstrate that CNs are nontoxic vehicles that can improve CpG uptake into tumor-associated inflammatory cells, leading to a more robust anti-tumor response.[194] Alizadeh et al.[195] evaluated the mechanism of cyclodextrin-based NP uptake into a murine glioma model. Using mixed in vitro culture systems, the authors demonstrated that cyclodextrin-based NP was preferentially taken up by BV2 and N9 microglia cells as compared to GL261 glioma cells. Fluorescent microscopy and flow cytometry analysis of intracranial GL261 gliomas confirmed these findings, and demonstrated a predominant cyclodextrin-based NP uptake by macrophages and microglia within and around the tumor site. In conclusion, these studies better characterize the cellular distribution of cyclodextrin-based NP in brain tumors, and demonstrate that macrophages and microglia could potentially be used as NP drug carriers into malignant brain tumors. Schneider et al.[171] recently examined a "double-punched" approach to overcome the escape of GBM cells from immune surveillance through an active specific immunization. The authors, using Newcastle disease virus-infected tumor cells and an antisense oligonucleotide against the TGF-β loaded in polybutyl cyanoacrylate NPs, demonstrated a significant decrease in plasma TGF-β_2 and an increase in antitumoral cytotoxicity.

In vitro study of PDT with targeted delivery systems is an alternative to current adjuvant therapy, and it is characterized by low morbidity and lack of susceptibility to the development of resistance. PDT involves the use and delivery of photosensitizers combined with local excitation by the appropriate wavelength of light, resulting in the production of ROS, which initiate apoptosis and cytotoxicity in many types of tumors. Moreover, PDT produces microvascular injury within treated neoplastic tissue, leading to inflammation and hypoxia. This therapeutic approach is more selective and less toxic than chemotherapy, because the drug is not activated until the light is delivered. The recent molecular targeting approach selectively detects cancer cells through molecular-recognition processes such as ligand–receptor or antibody–antigen interactions. PDT is an interesting approach for the treatment of malignant gliomas, resulting in a localized treatment approach with an important improvement in local control of tumors and a significantly improved survival rate.[165] In a recent study, Wang et al.[162] have demonstrated prominent photothermal selective damage of 426 targeted glioma cells. Molecular targeting, in this case, has been performed using CNs conjugated with monoclonal antibodies specific to CD133. PDT using the nanobiohybrid photocatalyst resulted in the destruction of over 80% of A172 glioma cells with high levels of IL-13R expression, whereas in the case of U87 cells characterized by lower antigen presentation, cytotoxicity at the same conditions reached a plateau of 50% and did not increase at higher photocatalyst concentrations. Moreover, no cytotoxicity was observed for normal human astrocytes, which are known not to express IL-13R.[196,197] Recently Tian et al.[198] showed the feasibility of encapsulating the alkylating agent, temozolomide [(3,4-dihydro-3-methyl-4-oxoimidazo [5,1-d]-as-tetrazine-8-carboxamide (TMZ)], into polybutylcyanoacrylate NPs by polymerization. Compared with TMZ solution, TMZ–polybutylcyanoacrylate NPs exhibited sustained release in vitro. Furthermore, based on the pattern of distribution in body organs, higher concentrations of TMZ can be detected in the brain after binding to polybutylcyanoacrylate NPs coated with polysorbate-80, which may be more useful for treating brain tumors. The prepared formulation may also reduce the toxicity of chemotherapy. Steiniger et al.[199] demonstrated in a murine GBM model, a statistically significant increase in survival time of GBM-bearing rats treated with doxorubicin bound to polysorbate-coated NPs, compared with the control groups treated with other doxorubicin formulations. More than 20% of the animals showed a long-term remission and no drug–NP complex neurotoxicity was observed. One interesting approach consists of coating an NP with polysorbate 80, which adsorbs apolipoproteins B and E, and allows receptor-mediated endocytosis by brain capillary endothelial cells. In these studies, 40% of the rats treated with doxorubicin-loaded NPs survived the duration of the study (6 months), with no evidence of residual tumor. Similarly, PEGylated doxorubicin-loaded solid lipid NPs can enhance delivery across the BBB after intravenous administration in rabbits.[200] Doxorubicin was present in the brain only after administration of the NP formulation and the extent of doxorubicin transport was dependent on the extent of PEG modification. Recently, a promising chemotherapeutic drug (SN-38) incorporated in micelles was compared with CPT-11 (irinotecan), a prodrug of SN-38, for the treatment of GBM in mice. The growth-inhibitory effects of the drug-loaded micelles were 34-fold to 444-fold more potent than those of CPT-11. In addition, when the drug was incorporated in the nanovectors, a significantly potent antitumor activity against an orthotopic glioblastoma multiforme xenograft and significantly longer survival rates than CPT-11 were observed.[201]

A new strategy to achieve selective drug delivery to tumor tissue is magnetic targeting. This approach has the advantage of enhancing the attraction of drug-loaded magnetic NPs in cancer cells by using an externally applied magnetic field.[125] Among the other polymer-derived drug delivery systems, the nanoconjugate Polycefin (based on polymalic acid) has been studied in animal models of human glioma, using intracranial injections of human cancer cells. Antiangiogenic results have been obtained in rats by injection, in vivo, of human glioma

U87MG xenografts.[202] A novel study aimed to examine the applicability of polyethyleneimine-modified magnetic NPs (GPEI) as a potential vascular drug/gene carrier to brain tumors. The obtained data show that the cationic magnetic NPs GPEI exhibit high cell penetration ability and low cell toxicity. In addition, GPEI could be magnetically captured in glioma lesions following clinically viable intracarotid administration. Furthermore, the extent of GPEI accumulation was 5.2-fold higher than that of G100 in the tumor lesions, but not in the contralateral normal brain, revealing higher target selectivity of cationic NPs.[203]

Polycefin conjugated with appropriate monoclonal antbodies specific to a tumor cell surface receptor may represent a potential new drug for glioma treatment. A recent in vitro and in vivo study in rats has demonstrated the efficacy of indomethacin-loaded nanocapsules, with a significant reduction of the mitotic index and the microvascular proliferation in implanted glioma.[86] Recently, Etame et al.[204] have been described the first demonstration of focal enhanced delivery of GNPs with therapeutic potential into the cerebral hemisphere using MR-guided focused ultrasound surgery in a rat model. The authors show the first direct evidence of localization of GNPs within the brain parenchyma suggesting BBB transgression. These results suggest a potential role for MR-guided focused ultrasound surgery in the delivery of GNPs with therapeutic potential into the CNS for targeting neurological disorders.

4.4 Recent patents in brain tumors

The development of new targeted chemical compounds as well as noninvasive targeted strategies is gaining greater attention. However, the principal targets of the following reported patents are represented by the ideal of overcoming the BBB because an efficient overcoming of the BBB could permit better delivery of pharmacological compounds. It is clear that the methods discussed in these patents are equipped for the passage of various therapeutic and/or diagnostic agents, able to treat various brain diseases including brain tumors. Besides, various applications can also increase the efficacy, safety, and efficiency of chemotherapeutic or irradiation therapies.[205]

4.4.1 Magnetic and metallic nanoparticles

Oxidative stress and interference with mitochondrial energy production by magnetic NPs lead to in vivo cytotoxicity, which has limited their clinical translation. To bypass this effect, thermosensitive magnetic NPs prepared by covering the NPs with a thermosensitive polymer with a critical temperature of 40–45 °C were used for hyperthermic treatment of tumors. The thermosensitive polymer shell could be ruptured at the site of action by applying an external magnetic field to increase the temperature of the inner magnetic core.[206] Intratumoral thermotherapy may be used in combination with any conventional therapy to amplify its effects. Because of the stability of the NP deposits the thermotherapy sessions may be repeated or combined with other therapies without any inherent limit. The Nano Therm® therapy, also termed magnetic fluid hyperthermia, combined with fractionated stereotactic radiotherapy, is a new local heat treatment of solid tumors (such as glioblastoma multiforme and prostate carcinoma). It consists of NanoTherm®, Nanoplan® and NanoActivator™ F100 (MagForce Nanotechnologies AG, Berlin, Germany). NanoTherm is a magnetofluid made of superparamagnetic iron oxide NPs which are colloidally dispersed in water with a high iron concentration. The iron oxide magnetite (Fe_3O_4) core is approximately 12 nm in diameter and coated with an aminosilane-type shell. Due to their aminosilane coating, these small magnets can be finely dispersed in water, forming a colloidal solution that is dispensable with a syringe. Once inside the alternating magnetic field applicator, NanoActivator are responsible for the production of warmth. Through this high frequency magnetic field, the NPs begin oscillating and warmth is produced from directly within the tumor tissue. Hence, the tumor cells are either directly destroyed or sensitized for the accompanying chemotherapy or radiotherapy, the final effect depending on the temperature reached and the length of treatment. NanoTherm® therapy has been used successfully to treat glioblastoma multiforme in a phase II trial.[207] In this trial combined treatment of fractionated stereotactic radiotherapy and NanoTherm® therapy was applied to 59 patients with recurrent GBM. The magnetic fluid MFL AS1 (NanoTherm AS1, $c = 112$ mg/ml) was injected directly into the tumor with the help of neuro-navigational guide surgery. A fiber-optic thermometry probe was installed into the tumor after injection of the NPs for a temperature measurement during the hyperthermia sessions, which were carried out in a NanoActivator F100. The median temperature measured within the tumor area during the hyperthermia was 51.2 °C. The median overall survival from diagnosis of the first tumor recurrence among the 59 patients with recurrent GBM was 13.4 months after primary tumor diagnosis, in comparison with the previous 23.2 months. Any serious side effect of the therapeutic approach was observed.

Magnetic NPs dispersible in lipid and aqueous phases were also structured. These new NPs have a core surrounded by oleic acid and stabilized by a pluronic acid. These lipophilic magnetic NPs containing doxorubicin were capable of efficient targeting of the drug deep into the cancerous brain tissue. A recent patent is characterized by an internal magnet to which the drug could be attached. The internal layer is coated by a polymeric layer of dextran or silicon; this in turn is coated with an extra layer of polysorbate 80 so as to create a space between the shell and the magnetic core. The therapeutic agents could be attached to the projection of the embedded functional group that is extended from the magnetic NPs into the space between the magnetic NPs and the outer polysorbate layer. This approach shows that the nano-magnets, surrounded by a layer of dextran or silicon, could attach the drug in a highly stable system covered by an outer polysorbate 80 layer and that it enables BBB penetration due to the effect of polysorbates.[208] Newly, functionalized magnetic NPs were produced to contain a functional group with selectivity to some types of brain cancer such as astrocytoma and glioma. Additionally, it was found that the ratio of diameter of the magnetic core to the whole functionalized NP could affect the permeation through the BBB and various tissues. Recently, novel magnetic iron oxides NPs were designed for brain tumor targeting. These magnetic iron oxides NPs were covered with sulfated glycosaminoglycan compounds such as heparin. The coating molecule was attached noncovalently to polyethyleneimene or low-molecular-weight protamine to which the antitumor agent is attached. This patent allows the antitumor agent to pass from the carotid arteries through the BBB with the assistance of the penetrating peptide and then to target the tumor tissue by the effect of an external magnetic field.[209]

Thanks to their unique physicochemical properties, GNPs classically enter into the cells with a nonspecific receptor-mediated endocytosis mechanism[210] A patent regarding GNPs relates to a new method for imaging tumors by detecting tumor biomarkers.[211] It may also be used for killing tumor cells. Electrical impedance tomography is an imaging technique in which an image of the conductivity of a part of a body is inferred from surface electrical measurements. Conducting electrodes attached to the skin of the subject and small alternating currents applied to some or all of the electrodes measure the resulting electrical potentials. By targeting NPs to a tumor and detecting the location of the NPs using the electrical impedance tomography or, preferably, a multifrequency electrical impedance tomography, a more accurate image of the tumor can be produced. The electrical impedance tomography can be used for detecting NPs to which are attached ligands such as antibodies or fragments thereof which are capable of attaching to a biomarker on a tumor. Examples of specific tumor biomarkers that may be targeted include carcinoembryonic antigen on the surface of colon cancer cells, HER-2 a surface biomarker for breast cancer, EGFR and mesenchymal–epithelial transition factor, both biomarkers for brain tumors. The core material for a metal NP can be gold, ferrous iron, silver, copper or platinum or combinations thereof, for example an alloy selected from Au/Ag, Au/Cu, Au/Ag/Cu, Au/Pt, Au/Fe, Au/Cu or Au/Fe/Cu. By coating theranoparticles with molecules such as antibody fragments bound by different linkers, the NPs can be made a sufficient size to exploit the greater leakiness of the tumor blood vessels, thereby differentially targeting theranoparticles to the tumor. This selectivity can be augmented by warming with microwaves or alternating magnetic fields focused on the location of the tumor. These allow deeper interstitial penetration of a tumor mass. In addition to antibody or antibody fragment ligands, the NPs may have anticancer drugs, for example doxorubicin and/or cytotoxic compounds attached to them in order to treat the tumor to which they become attached or entrapped within.

A recent invention[212] provides methods of using metal NPs to enhance the dose and effectiveness of X-rays or of other kinds of radiation in therapeutic regimens of ablating a target tissue, such as tumor. The metal NPs can be administered intravenously, intra-arterially, or locally to achieve specific loading in and around the target tissue. The metal NPs can also be linked to chemical and/or biochemical moieties, which bind specifically to the target tissue so as to treat solid tumors such as carcinomas, brain tumors, melanomas, lymphomas, plasmocytomas, sarcomas, and thymomas. Metal NPs suitable for use in radiation therapy are composed of a metal core and typically a surface layer surrounding the metal core. Metals that can be used to form NPs suitable for enhancing radiation effects are heavy metals, or metal with a high Z number, including gold, silver, platinum, palladium, cobalt, iron, copper, tin, tantalum, vanadium, molybdenum, tungsten, osmium, iridium, rhenium, hafnium, thallium, lead, bismuth, gadolinium, dysprosium, holmium, and uranium. The preferred metal is gold. The metal core can consist of one metal, or it can be a mixture or an ordered, concentric layering of such metals, or a combination of mixtures and layers of such metals. The shell or surface layer material typically surrounding the metal can be molecules containing sulfur, phosphorus or amines, e.g. phosphines, phenanthrolines, silanes and organo-thiols. Other surface layer materials suitable for use in accordance with the present invention include synthetic polymers, proteins, antibodies, antibody fragments, peptides, nucleic acids, carbohydrates, lipids, drugs, and other compounds.

A recent patent developed divalent-metal coated NPs for delivery of compositions into the CNS by nasal insufflation.[213] The compositions and methods of the disclosure particularly target the divalent metal transporter expressed on olfactory nerve terminals to transport divalent cation-coated or cation-containing NPs to all regions of brain. It has been found that such divalent cation-containing NPs, including those NPs comprising manganese have affinity for the metal transport receptor proteins. Although this receptor has particular affinity for manganese, it is contemplated that other divalent ions, such as magnesium and calcium, may also be bound to such receptors leading to transport of the NPs into the intracellular cytoplasm. Nanoparticles have been developed, therefore, as vehicles for parenteral delivery of genes, oligonucleotides, including but not limited to siRNA, and other small molecule drugs, into the brain by nasal insufflation.

In a recent patent[214] modified GNPs enable a noninvasive, real time, targeted cancer imaging–therapeutic in one step. After reaching the cancer targets, the designed targeted GNPs significantly enhance conventional treatment modalities at the cellular level, and the GNPs of this invention are modified to be bound to a positron emission tomography tracer.

4.4.2 Polymeric nanoparticles

Polymeric NPs, namely nanospheres and nanocapsules, show greater stability in biological fluids and during the enzymatic metabolism. The use of biodegradable, biocompatible polymers such as PCL and ε-caprolactone, to synthesize a polymer scaffold into which the NPs are dispersed further extends drug release over several months, as the slow degradation of the scaffold allows for prolonged, controlled release of drug-loaded NPs, removing the need for daily oral intake and thereby enhancing the patient's quality of life. The availability of numerous polymer fabrication techniques reported enables researchers to manipulate the physicochemical and physicomechanical properties of the material to obtain optimum drug-release kinetics from innovative delivery systems. This application[215] relates to a polymeric pharmaceutical dosage form for the delivery, in use, of at least one pharmaceutical composition in a rate-modulated and site-specific manner. The dosage form comprises a biodegradable, polymeric, scaffold loaded with at least one active pharmaceutical compound. The polymer or polymers making up the scaffold degrade in a human or animal body in response to or in the absence of specific biological stimuli and, on degradation, release the pharmaceutical compound in an area where said stimuli are encountered.

Given that the chemotherapic TMZ is toxic, its therapeutic dosages are limited by severe side effects. A targeted delivery can improve the efficiency of TMZ and reduce healthy tissue toxicity. Recently, multifunctional targetable nanoconjugates of TMZ hydrazide have been synthesized using a poly(L-malic acid) platform, which contained a targeting monoclonal antibody to transferrin receptor, trileucine for pH-dependent endosomal membrane disruption, and PEG for protection. The strongest reduction of human brain cancer cell viability was obtained using versions of TMZ nanoconjugates containing trileucine and anti-transferrin receptor antibody.[216] The release of chemotherapeutic agents from implantable drug–polymer carrier systems intended for local delivery can be further delayed and modulated by embedding drug-loaded NPs within a polymer matrix in the place of pure drug. The combined unique hydration and swelling dynamics of each system gives rise to higher-order drug-release kinetics and drug-modulation effect compared with a matrix system loaded with pure drug, rendering the composite system more suitable for long-term drug delivery.

An interesting patent is characterized by the combination of two, three, or more agents useful in treating a patient with a neoplasm, as well as methods for identifying combinations of compounds potentially useful in treating a patient with a neoplasm. Combinations can include cerivastatin, irinotecan, lovastatin, topotecan, simvastatin conjugated with adefovir dipivoxil, auranofin, epirubicin, idebenone.[217] An original patent to target drugs to brain neurons[218] has been applied on some selected types of neural cells and/or brain neurons, such as cerebellar Purkinje cells, raphe nucleus neurons, cerebral cortex neurons, hypothalamus neurons, thalamus neurons, and brainstem neurons. The drug can be administered directly by injecting it into the cerebrospinal fluid by lumbar puncture. In this conventional drug delivery system, after processing such as encapsulation of a small amount of a drug in small capsules (drug carriers) that are tens of nanometers across has been performed, the drug is selectively delivered to the target cells such as those of cancerous tissue by exploiting interactive mechanisms such as the antigen–antibody reaction. Examples of carriers used include water-soluble polymers, nanospheres, liposomes, and polymeric micelles where polymers with heterogeneous structure are aggregated. This drug delivery system is a method in which the drug is coupled to the drug carrier and administered into a blood vessel to deliver it to the affected area via the circulatory system, and the drug is made to act on the targeted cells through chemical or physical mechanisms. Although this patent was born to target brain neurons in particular to modulate the concentration of CNS-active molecules and/or neurotransmitters in patients affected by neurodegenerative diseases, this same kind of approach may be applied in neuro-oncology to treat cerebral gliomas and overcome the actual problems of the toxic effects of systemic chemotherapy, overcoming the BBB, drug delivery to selective cancer cells, and cell targeting.

Another recent patent[219] has been developed through the simultaneous grafting of polymethacrylic acid grafted starch and NP formation in an aqueous medium. The new pH-responsive NPs possessed useful properties for controlled drug delivery.

4.4.3 Lipid nanoparticles

Nanoparticles comprising lipids, such as brain lipids, are used to encapsulate drugs and to facilitate drug delivery to the brain.[220]

The liposome-incorporated entity may be completely or partially located within the bilayer membrane of the liposome, or associated with the exterior surface of the liposome membrane. In a recent patent, NP composition was characterized by brain lipids, supplemental lipids, PEG-conjugated lipids and drugs.[221] The NPs are generally made by dissolving one or more drugs and brain lipids in chloroform. In an experimental application, NPs containing brain-derived lipids were formulated encapsulating 6-coumarin, which resulted in a fivefold increase in drug levels inside the brain compared with controls. Furthermore, hydrophobic ion pairing agents were found to be efficient for inclusion of BBB-impermeable proteins into lipophilic core NPs, hence rendering them BBB permeable. Anionic proteins could bind cationic hydrophobic ion pairing agents such as cetrimonium bromide, leading to efficient encapsulation of the protein into the lipophilic poly-butyl-cyanoacrylate NPs.[222] In a patent application, Micklus et al. structured an immunoliposome, capable of targeting pharmacological compounds to the brain. Liposomes are coupled to a monoclonal antibody binding fragment such as Fab, $F(ab')_2$, Fab' or a single-chain polypeptide antibody that binds to a receptor molecule present on the vascular endothelial cells of the mammalian BBB. The receptor is preferably from the brain peptide transport system, such as the transferring receptor, or insulin receptor, insulin-like growth factor-1 or -2 receptor.[223] Another patent[224] demonstrates that substituted ammonium and polyanions are useful for loading and retaining entities inside liposomes. Accordingly, the present invention provides methods and liposome compositions that are useful for the delivery of a variety of entities, especially therapeutic entities, that is, entities useful in the diagnosis, prognosis, testing, screening, treatment, or prevention of an undesirable condition. An interesting application[225] provides a method of CED of drugs to cross the BBB. In particular, it related to a method using self-diffusion delivery of drug in the extracellular space of the brain to make the drug reach the target brain tissue and produce an effect. Cytidine-5′-diphosphate choline is a micromolecule that finds it difficult to cross the BBB because of its strong polarity. Because the neuroprotection of citicoline is dose-dependent, the amount of cytidine-5′-diphosphate choline absorbed by brain tissue is one of the key factors in thereapeutic efficacy. The intake of citicoline is improved usually by increasing the dosage of drugs and enhancing the permeability of BBB via liposomes. This application can reduce delivery time and dose of drugs, relieve injection pressure, decrease damage to normal brain tissue, and observably reduce the cost of treatment.

A recent patent[226] relates to a highly efficient artificial LDL carrier system for the targeted delivery of therapeutic agents across the BBB. In particular, this invention relates to artificial LDL particles comprised of three lipid elements: phosphatidyl choline, fatty-acyl-cholesterol esters, and at least one apolipoprotein. The invention provides a process for conjugating hydrophilic therapeutic agents with cholesterol to facilitate incorporation of the conjugated therapeutic agent into an artificial LDL particle. In a preferred format, the present invention provides cholesterol-conjugated adriamycin and tetracycline.

Another patent[227] provides a novel synthetic LDL NP comprising a lipid moiety and a synthetic chimeric peptide wherein the lipid moiety forms a particle that is c.10–30 nm in size and the synthetic chimeric peptide comprises an amphipathic α-helix and an LDL receptor. The LDL receptor ligand is apoB100, a 514-kDa glycoprotein on the surface of LDL. Various studies suggest that tumors have a high requirement for LDL. The increased import of LDL into cancerous cells is thought to be the result of elevated LDL receptors in these tumors. Comparative studies of normal and malignant brain tissues have shown a high propensity of LDL receptors to be associated with malignant or rapidly growing brain cells and tissues.[228] These findings suggest that malignant brain tumors exhibit increased expression of LDL receptors because of their increased requirement for cholesterol.

4.4.4 Peptide nanoparticles

The interaction of peptides and proteins with the cell membrane results in their penetration into the cell or the formation of pores within the cell membrane.[229] A recent application[230] provides a conjugate comprising a transmembrane module (TPU), a nuclear localization sequence (NLS) and a signaling and/or drug-carrying module. The use of gadolinium (Gd^{3+}) contrast agents is limited to the extracellular space. To overcome this problem, a transmembrane transport of Gd^{3+} an amphiphilic transport peptide of human origin (TPU) that contains a similar peptide sequence to that of the homeodomain of Antennapedia, was realized. This similar peptide sequence was chosen to minimize the risk of immunization reactions. This transport peptide is part of a modularly constructed CNN–Gd[3k] complex and is cleavably covalently linked to the nuclear localization sequence of SV40T–antigen via a disulfide bond. After cleavage of the disulfide bond, the nuclear localization sequence becomes the terminal part of the conjugate and can be recognized by the cytoplasmic receptor (importin α). After binding of the complex to a second cytoplasmic receptor (importin β) the entire complex Gd^{3+}–nuclear localization sequence–importin α–importin β is delivered to the nucleus. Using MRI, Gd^{3+} was detected within DU-145 prostate cancer cells after only 10 min.

4.4.5 Theranostic nanoparticles

The multifunctionality of nanovehicles creating nanotheranostic platforms offers a number of advantages over conventional agents and are powerful tools for imaging and treatment of cancer. These include targeting to a diseased site, thereby minimizing systemic toxicity; the ability to solubilize hydrophobic or labile drugs, leading to improved pharmacokinetics; and their potential to image, treat, and predict the therapeutic effects.

In various embodiments, the surface of the NP is modified or functionalized with at least a portion of an isolated cellular membrane, such as an isolated plasma membrane. Often, the NP is a lipid particle or a liposome that contains a lipid layer. In other settings the NP is fabricated as a porous particle (e.g. a porous silicon or a porous silica) or a multistage object. In addition, the NP contains at least one active agent, such as therapeutic and/or imaging agents. This invention provides targeted nanoplex molecules that carry multimodality imaging reporters together with target enzyme inhibitors such as siRNAs and target prodrug enzymes, that are useful for theranostic imaging. The nanoplex molecules of the present invention provide a platform technology toward many cancer subtypes and alternative therapeutic targets. Downregulation of specific pathways using targeted enzyme inhibitors further provides unique opportunities to target cancer cells selectively while sparing normal tissue. Such a nanoplex molecule platform also has the ability to deliver multiple siRNA enzyme inhibitors. The strategy described herein can be useful to downregulate multidrug resistance pathways, or to repair enzymes with the goal of increasing the efficacy, safety, and efficiency of chemotherapeutic or irradiation therapies. Small interfering RNA-mediated silencing of specific targets has significant potential in cancer therapy to downregulate pathways that are upregulated in cancer cells but not in normal tissue to achieve cancer cell-specific treatment. Similarly, prodrug enzyme therapy, where a drug-activating enzyme delivered to the tumor converts a nontoxic prodrug to a cytotoxic drug, is being actively investigated to minimize normal tissue damage. A combination of both strategies can be exploited to enhance the effect of conventional chemotherapy against cancer cells and minimize damage to normal tissue. Imaging can play a key role in several aspects of such a treatment. The ability to image the delivery of the siRNA and the prodrug-activating enzyme within the tumor would ascertain effective delivery.

4.5 Nanotechnology

Nanotoxicology evaluates the interactions of NPs with biological systems and the relationship between the physical and chemical properties of NPs with the induction of toxic biological responses. Currently, a complete evaluation of the size, shape, composition, and aggregation-dependent interactions of NPs with biological systems is lacking, so it is unclear whether the exposure of humans, animals, and plants to engineered nanostructures could produce harmful biological responses.

Nanotechnology has been considered by some as a risk to human health.[231] Although its benefits are widely publicized, concerns on the potential effects and safety of its widespread use in consumer and industrial products is becoming frequent. While developing therapeutics, attention should be given to toxicity. It is hard to argue on this matter because of the limited information available. Some have drawn an analogy between high-aspect-ratio NPs and asbestos fibers. In the UK, the Prince of Wales has requested advice on nanotechnology from the Royal Society, whereas Greenpeace and the Canadian Action Group on Erosion, Technology, and Concentration have called for a moratorium on the use of NPs until the toxicological issues have been resolved. Although some concerns may be ill-founded, it remains true that the toxicology of many nanomaterials has not yet been fully evaluated. To address this issue, some companies are participating in the European Nanosafe consortium, which is starting to evaluate the possible risks presented by nanomaterials. In the USA, the Center for Biological and Environmental Nanotechnology at Rice University has begun an investigation of two popular nanomaterial systems: carbon nanotubes and titanium dioxide.[232] The NPs own electronic, optical, and magnetic properties that are related to their physical dimensions, and their breakdown could lead to a unique toxic effect that is difficult to predict. Although targeted NPs have emerged as one strategy to overcome the lack of specificity of conventional chemotherapy, there are other potential risks and challenges associated with this novel strategy. Some cancer cell types would develop drug resistance, rendering drugs released from the targeted

NPs ineffective. Also the targeted NPs might change the stability, solubility, and pharmacokinetic properties of the carried drugs. The shelf life, aggregation, leakage, and toxicity of materials used to make NPs are other limitations for their use. Some materials used to make NPs show low toxicity, but degrade quickly and do not circulate in tissues long enough for sustained drug/gene delivery. On the other hand, other materials, such as carbon nanotubes and quantum dots are durable and can persist in the body for weeks, months, or even years, making them potentially toxic and limiting their use for repeated treatments. The surfaces of NPs, also, are involved in many catalytic and oxidative processes that may be potentially cytotoxic. Some NPs contain metals or compounds with known toxicity, so the breakdown of these materials could elicit similar toxic responses to the components themselves. Nanomaterials can enter the human body through several ports. Accidental or involuntary contact during production or use is most likely to occur via the lungs, from which a rapid translocation is possible to other vital organs through the bloodstream. On the cellular level, an ability to act as a gene vector has been demonstrated for NPs. Carbon black NPs have been implicated in interfering with cell signaling. There is work that demonstrates uses of DNA for the size separation of carbon nanotubes. The DNA strand just wraps around it if the tube diameter is right.[233] Though excellent for the purposes of separation, this tendency raises some concerns over the consequences of carbon nanotubes entering the human body. Coated or uncoated NPs have a tendency to accumulate in the liver in varying amounts. Therefore, a detailed mechanism for elimination from the body/ metabolization needs to be properly addressed. Moreover, while formulating NPs, an important aspect is to minimize the batch-to-batch variations; the synthetic yield and the drug-loading effectiveness must also be boosted to warrant practical utility of the NPs. In addition, the mechanism of endocytosis and degradation pathways needs to be addressed because these are still poorly understood, despite their primary importance for clinical transition.[234] Furthermore, there is an urgent need for the development of safety guidelines regarding the environmental effects and the potential effects on the health of people manufacturing the NPs. Despite these concerns, the most exciting prospect of nanocarriers is the near limitless possibilities for treatment strategies. The versatility of formulation, colloidal size, biocompatibility, and sustained-release properties of NPs have already been accepted with growing interest for a wide range of applications. However, the recent strategy gaining widespread interest and attention is that of "functionalized NPs". Methods are being devised to tailor the surface characteristics of NPs to achieve specific ligand-mediated targeting of therapeutic and imaging agents. Optimization of these techniques, coupled with the considerable progress made in the field of NP characterization and a better understanding of the in vivo behavior of NPs have raised hopes for the successful development of commercial products based on targeted NPs for use in therapy and imaging.

References

1. Juliano, R.L. (1978) Drug delivery systems: a brief review. *Can. J. Physiol. Pharmacol.*, **56**: 683–90.
2. Parveen, S., and Sahoo, S.K. (2006) Nanomedicine: clinical applications of polyethylene glycol conjugated proteins and drugs. *Clin. Pharmacokinet.*, **45**: 965–88.
3. Lee, H.J. (2002) Protein drug oral delivery: the recent progress. *Arch. Pharm. Res.*, **25**: 572–84.

4. Matsudai, M., and Hunt, G. (2005) Nanotechnology and public health. *Nippon Koshu Eisei Zasshi*, **52**: 923–7.
5. Chan, W.C. (2006) Bionanotechnology progress and advances. *Biol. Blood Marrow Transplant*, **12**: 87–91.
6. Sahoo, S.K., Parveen, S., and Panda, J.J. (2007) The present and future of nanotechnology in human health care. *Nanomedicine*, **3**: 20–31.
7. Byrne, J.D., Betancourt, T., and Brannon-Peppas, L. (2008) Active targeting schemes for nanoparticle systems in cancer therapeutics. *Adv. Drug Deliv. Rev.*, **60**: 1615–26.
8. Heath, T.D., Fraley, R.T., and Papahdjopoulos, D. (1980) Antibody targeting of liposomes: cell specificity obtained by conjugation of $F(ab')_2$ to vesicle surface. *Science*, **210**: 539–41.
9. Ferreira, E., Potier, E., Logeart-Avramoglou, D., Salomskaite-Davalgiene, S., Mir, L.M., and Petite, H. (2008) Optimization of a gene electrotransfer method for mesenchymal stem cell transfection. *Gene Ther.*, **15**: 537–44.
10. Arbab, A.S., Bashaw, L.A., Miller, B.R., Jordan, E.K., Lewis, B.K., et al. (2003) Characterization of biophysical and metabolic properties of cells labeled with superparamagnetic iron oxide nanoparticles and transfection agent for cellular MR imaging. *Radiology*, **229**: 838–46.
11. Gershon, H., Ghirlando, R., Guttman, S.B., and Minsky, A. (1993) Mode of formation and structural features of DNA-cationic liposome complexes used for transfection. *Biochemistry*, **32**: 7143–51.
12. Musyanovych, A., Schmitz-Wienke, J., Mailander, V., Walther, P., and Landfester, K. (2008) Preparation of biodegradable polymer nanoparticles by miniemulsion

technique and their cell interactions. *Macromol. Biosci.*, **8**: 127–39.

13. Mailander, V., Lorenz, M.R., Holzapfel, V., Musyanovych, A., Fuchs, K., et al. (2008) Carboxylated superparamagnetic iron oxide particles label cells intracellularly without transfection agents. *Mol. Imaging Biol.*, **10**: 138–46.
14. Lorenz, M.R., Holzapfel, V., Musyanovych, A., Nothelfer, K., Walther, P., et al. (2006) Uptake of functionalized, fluorescent-labeled polymeric particles in different cell lines and stem cells. *Biomaterials*, **27**: 2820–8.
15. Gratton, S.E., Ropp, P.A., Pohlhaus, P.D., Luft, J.C., Madden, V.J., et al. (2008) The effect of particle design on cellular internalization pathways. *Proc. Natl. Acad. Sci. USA*, **105**: 11613–18.
16. Panyam, J., and Labhasetwar, V. (2003) Dynamics of endocytosis and exocytosis of poly(D,L-lactide-co-glycolide) nanoparticles in vascular smooth muscle cells. *Pharm. Res.*, **20**: 212–20.
17. Panyam, J., Zhou, W.Z., Prabha, S., Sahoo, S.K., and Labhasetwar, V. (2002) Rapid endo-lysosomal escape of poly(dl-lactide-co-glycolide) nanoparticles: implications for drug and gene delivery. *Faseb J.*, **16**: 1217–26.
18. Davda, J., and Labhasetwar, V. (2002) Characterization of nanoparticle uptake by endothelial cells. *Int. J. Pharm.*, **233**: 51–9.
19. Roger, M., Clavreul, A., Venier-Julienne, M.C., Passirani, C., Sindji, L., et al. (2010) Mesenchymal stem cells as cellular vehicles for delivery of nanoparticles to brain tumors. *Biomaterials*, **31**: 8393–401.
20. Banerji, S.K., and Hayes, M.A. (2007) Examination of nonendocytotic bulk transport of nanoparticles across phospholipid membranes. *Langmuir*, **23**: 3305–13.

21. Rejman, J., Oberle, V., Zuhorn, I.S., and Hoekstra, D. (2004) Size-dependent internalization of particles via the pathways of clathrin- and caveolae-mediated endocytosis. *Biochem. J.*, **377**(Pt 1): 159–69.
22. Schulze, E., Ferrucci, Jr J.T., Poss, K., Lapointe, L., Bogdanova, A., and Weissleder, R. (2003) Cellular uptake and trafficking of a prototypical magnetic iron oxide label in vitro. *Invest. Radiol.*, **30**: 604–10.
23. Conner, S.D., and Schmid, S.L. (2003) Regulated portals of entry into the cell. *Nature*, **422**(6927): 37–44.
24. Paillard, A., Hindre, F., Vignes-Colombeix, C., Benoit, J.P., and Garcion, E. (2010) The importance of endo-lysosomal escape with lipid nanocapsules for drug subcellular bioavailability. *Biomaterials*, **31**: 7542–54.
25. Jain, R.A. (2000) The manufacturing techniques of various drug loaded biodegradable poly(lactide-co-glycolide) (PLGA) devices. *Biomaterials*, **21**: 2475–90.
26. Prabha, S., and Labhasetwar, V. (2004) Critical determinants in PLGA/PLA nanoparticle-mediated gene expression. *Pharm. Res.*, **21**: 354–64.
27. Labhasetwar, V., Bonadio, J., Goldstein, S., Chen, W., and Levy, R.J. (1998) A DNA controlledrelease coating for gene transfer: transfection in skeletal and cardiac muscle. *J. Pharm. Sci.*, **87**: 1347–50.
28. Duncan, R. (2006) Polymer conjugates as anticancer nanomedicines. *Nat. Rev. Cancer*, **6**: 688–701.
29. Fonseca, C., Simoes, S., and Gaspar, R. (2002) Paclitaxel-loaded PLGA nanoparticles: preparation, physiochemical characterization and in vitro anti-tumoral activity. *J. Control Release*, **83**: 273–86.
30. Avgoustakis, K., Beletsi, A., Panagi, Z., Klepetsanis, P., Karydas, A.G., and Ithakissios, D.S. (2002) PLGA-mPEG nanoparticles of cisplatin: in vitro nanoparticle

degradation, in vitro drug release and in vivo drug residence in blood properties. *J. Control Release*, **79**: 123–35.
31. Gao, H., Wang, Y.N., Fan, Y.G., and Ma, J.B. (2005) Synthesis of a biodegradable tadpole-shaped polymer via the coupling reaction of polylactide onto mono(6-(2-aminoethyl)amino-6-deoxy)-β-cyclodextrin and its properties as the new carrier of protein delivery system. *J. Control Release*, **107**: 158–73.
32. Koo, Y.L., Reddy, G.R., Bhojani, M., Schneider, R., Philbert, M.A., et al. (2006) Brain cancer diagnosis and therapy with nanoplatforms. *Adv. Drug Deliv. Rev.*, **58**: 1556–7.
33. Pinhassi, R.I., Assaraf, Y.G., Farber, S., Stark, M., Ickowicz, D., et al. (2010) Arabinogalactan-folic acid-drug conjugate for targeted delivery and target-activated release of anticancer drugs to folate receptor-overexpressing cells. *Biomacromolecules*, **11**: 294–303.
34. Rice, J.R., Gerberich, J.L., Nowotnik, D.P., and Howell, S.B. (2006) Preclinical efficacy and pharmacokinetics of AP5346, a novel diaminocyclohexane-platinum tumor targeting drug delivery system. *Clin. Cancer Res.*, **1**: 2248–54.
35. Matsumura, Y. (2008) Poly (amino acid) micelle nanocarriers in preclinical and clinical studies. *Adv. Drug Deliv. Rev.*, **60**: 899–914.
36. Homsi, K., Simon, G.R., Garrett, C.R., Springett, G., and De Centi, R. (2007) Phase I trial of poly-L-glutamate comptothecin (CT-2106) administered weekly in patients with advanced solid malignancies. *Clin. Cancer Res.*, **13**: 5855–61.
37. Boddy, A.V., Plummer, E.R., Todd, R., Sludden, J., Griffin, M., et al. (2005) A phase I and pharmacokinetic study of paclitaxel poliglumex (XYOTAX), investigating

both 3-weekly and 2-weekly schedules. *Clin. Cancer Res.*, **11**: 7834–40.

38. Savic, R., Eisenberg, A., and Maysinger, D. (2006) Block copolymer micelles as delivery vehicles of hydrophobic drugs: micelle-cell interactions. *J. Drug Target*, **14**: 343–55.
39. Shen, Y., and Jiasheng, T. (2009) Synthesis and characterization of low molecular weight hyaluronic acid-based cationic micelles for efficient siRNA delivery. *Carbohydrate Polym.*, **77**: 95–104.
40. Kim, D.W., Kim, S.Y., Kim, H.K., Kim, S.W., Shin, S.W., et al. (2007) Multicenter phase II trial of Genexol-PM, a novel Cremophor-free, polymeric micelle formulation of paclitaxel, with cisplatin in patients with advanced non-small-cell lung cancer. *Ann. Oncol.*, **18**: 2009–14.
41. Kenmotsu, H., Yasunaga, M., Goto, K., Nagano, T., Kuroda, J., et al. (2010) The antitumor activity of NK012, an SN-38-incorporating micelle, in combination with bevacizumab against lung cancer xenografts. *Cancer*, **116**: 4597–604.
42. Bey, E.A., Bentle, M.S., Reinicke, K.E., Dong, Y., Yang, C.R., et al. (2007) An NQO1- and PARP-1-mediated cell death pathway induced in non-small-cell lung cancer cells by β-lapachone. *Proc. Natl. Acad. Sci. USA*, **104**: 11832–7.
43. Blanco, E., Bey, E.A., Khemtong, C., Yang, S.G., Setti-Guthi, J., et al. (2010) β-Lapachone micellar nanotherapeutics for non-small cell lung cancer therapy. *Cancer Res.*, **70**: 3896–904.
44. Khong, T., and Spencer, A. (2011) Targeting heat shock protein 90 induces apoptosis anf inhibits critical survival and proliferation pathways in multiple myeloma. *Mol. Cancer Ther.*, **10**: 1909–17.

45. Richardson, P.G., Ghana-khan, A.A., Alsina, M., Albitar, M., Berman, D., et al. (2010) Tanespimycin monotherapy in relapsed multiple myeloma: results of a phase 1 dose-escalation study. *Br. J. Haematol.*, **150**: 438–45.
46. Misra, A., Ganesh, S., and Shahiwala, A. (2003) Drug delivery to the central nervous system: a review. *J. Pharm. Pharmaceut. Sci.*, **6**: 252–73.
47. Needham, D., Anyarambhatla, G., Kong, G., and Dewhirst, M.W. (2000) A new temperature-sensitive liposome for use with mild hyperthermia: characterization and testing in a human tumor xenograft model. *Cancer Res.*, **60**: 1997–201.
48. Vail, D.M., Amantea, M.A., Colbern, G.T., Martin, F.J., Hilger, R.A., and Working, P.K. (2004) Pegylated liposomal doxorubicin: proof of principle using preclinical animal models and pharmacokinetic studies. *Semin. Oncol.*, **31**: 16–35.
49. Lorusso, D., Ferrandina, G., Lo Voi, R., Fagotti, A., and Scambia, G. (2004) Role of pegylated liposomal doxorubicin (PLD) in epithelial ovarian cancer. *J. Chemother.*, **4**: 98–103.
50. Kuesters, G.M., and Campbell, R.B. (2010) Conjugation of bevacizumab to cationic liposomes enhances their tumor-targeting potential. *Nanomedicine*, **5**: 181–92.
51. Zamboni, W.C. (2008) Concept and clinical evaluation of carrier-mediated anticancer agents. *Oncologist*, **13**: 248–60.
52. Ciuffreda, L., Del Bufalo, D., Desideri, M., Di Sanza, C., Stoppacciaro, A., et al. (2009) Growth-inhibitory and antiangiogenic activity of the MEK inhibitor PD0325901 in malignant melanoma with or without BRAF mutations. *Neoplasia*, **11**: 720–31.
53. Kang, S.H., Cho, H.J., Shim, G., Lee, S., Kim, S.H., et al. (2011) Cationic liposomal co-delivery of small

interfering RNA and a MEK inhibitor for enhanced anticancer efficacy. *Pharm. Res.*, **28**: 3069–78.

54. Dritschilo, A., Huang, C.H., Rudin, C.M., Marshall, J., Collins, B., et al. (2006) Phase I study of liposome-encapsulated c-raf antisense oligodeoxyribonucleotide infusion in combination with radiation therapy in patients with advanced malignancies. *Clin. Cancer Res.*, **12**: 1251–9.
55. Ressler, S., Mlineritsch, B., and Greil, R. (2011) Zoledronic acid for adjuvant use in patients with breast cancer. *Expert Rev. Anticancer Ther.*, **11**: 333–49.
56. Marra, M., Salzano, G., Leonetti, C., Tassone, P., Scarsella, M., et al. (2011) Nanotechnologies to use bisphosphonates as potent anticancer agents: the effects of zoledronic acid encapsulated into liposomes. *Nanomedicine*, **403**: 292–7.
57. Chithrani, B.D., Ghazani, A.A., and Chan, W.C. (2006) Determining the size and shape dependence of gold nanoparticle uptake into mammalian cells. *Nano Lett.*, **6**: 662–8.
58. Libutti, S.K., Paciotti, G.F., Myer, L., Haynes, R., Gannon, W., et al. (2009) Results of a completed phase I clinical trial of CYT-6091: a pegylated colloidal gold-TNF. *Nanomedicine*, **27**: 3586.
59. Hirsch, L., Stafford, R., Bankson, J., Sershen, S., Rivera, B., et al. (2003) Nanoshell-mediated near-infrared thermal therapy of tumors under magnetic resonance guidance. *Proc. Natl. Acad. Sci. USA*, **100**: 13549–54.
60. Hede, S., and Huilgol, N. (2006) Nano: the new nemesis of cancer. *J. Cancer Res. Ther.*, **2**: 186–95.
61. Kattumuri, V., Katti, K., Bhaskaran, S., Boote, E.J., Casteel, S.W., et al. (2007) Gum arabic as a phytochemical construct for the stabilization of gold nanoparticles: in

vivo pharmacokinetics and X-ray-contrast-imaging studies. *Small*, **3**: 333–41.

62. Shi, X., Wang, S., Meshinchi, S., Van Antwerp, M.E., Bi, X., et al. (2007) Dendrimer-entrapped gold nanoparticles as a platform for cancer-cell targeting and imaging. *Small*, **3**: 1245–52.
63. Bhattacharya, R., and Mukherjee, P. (2008) Biological properties of "naked" metal nanoparticles. *Adv. Drug Deliv. Rev.*, **60**: 1289–306.
64. Lara, H.H., Ayala-Nunez, N.V., Ixtepan-Turrent, L., and Rodriguez-Padilla, C. (2010) Mode of antiviral action of silver nanoparticles against HIV-1. *J. Nanobiotechnol.*, **8**: 1.
65. Sheikpranbabu, S., Kalishwaralal, K., Venkataraman, D., Eom, S.H., Park, J., and Gurunathan, S. (2009) Silver nanoparticles inhibit VEGF-and IL-1 -induced vascular permeability via Src dependent pathway in porcine retinal endothelial cells. *J. Nanobiotechnol.*, **7**: 8.
66. Sriram, M.I., Kanth, S.B.M., Kalishwaralal, K., and Gurunathan, S. (2010) Antitumor activity of silver nanoparticles in Dalton's lymphoma ascites tumor models. *Int. J. Nanomed.*, **5**: 753–62.
67. Zhang, X., Kono, H., Liu, Z., Nishimoto, S., Tryk, D.A., et al. (2007) A transparent and photo-patternable superhydrophobic film. *Chem. Commun.*, **46**: 4949–51.
68. Theron, J., Walker, J.A., and Cloete, T.E. (2008) Nanotechnology and water treatment: applications and emerging opportunities. *Crit. Rev. Microbiol.*, **34**: 43–69.
69. Yamaguchi, S., Kobayashi, H., Narita, T., Kanehira, K., Sonezaki, S., et al. (2010) Novel photodynamic therapy using water-dispersed TiO_2-polyethylene glycol compound: evaluation of antitumor effect on glioma

cells and spheroids in vitro. *Photochem. Photobiol.*, **86**: 964–71.

70. Seo, J.W., Chung, H., Kim, M.Y., Lee, J., Choi, I.H., and Cheon, J. (2007) Development of water-soluble single-crystalline TiO_2 nanoparticles for photocatalytic cancer-cell treatment. *Small*, **3**: 850–3.
71. Rozhkova, E.A., Ulasov, I., Lai, B., Dimitrijevic, N.M., Lesniak, M.S., and Rajh, T. (2009) A high-performance nanobio phocatalyst for targeted cancer-cell therapy. *Nano Lett.*, **9**: 3337–42.
72. Rajh, T., Dimitrijevic, N.M., and Rozhkova, E.A. (2011) Titanium dioxide nanoparticles in advanced imaging and nanotherapeutics. *Methods Mol. Biol.*, **726**: 63–75.
73. Lopez, T., Figueras, F., Manjarrez, J., Bustos, J., Alvarez, M., et al. (2010) Catalytic nanomedicine: a new field in antitumor treatment using supported platinum nanoparticles. In vitro DNA degradation and in vivo tests with C6 animal model on Wistar rats. *Eur. J. Med. Chem.*, **45**: 1982–90.
74. Dimitrijevic, N.M., Rozhkova EA, Rajh T (2009) Dynamics of localized charges in dopamine-modified TiO_2 and their effect on the formation of reactive oxygen species. *J. Am. Chem. Soc.*, **131**: 2893–9.
75. Koo, Y.L., Reddy, G.R., Bhojani, M., Schneider, R., Philbert, M.A., et al. (2006) Brain cancer diagnosis and therapy with nanoplatforms. *Adv. Drug Deliv. Rev.*, **58**: 1556–7.
76. Wang, Y.X., Hussain, S.M., and Krestin, G.P. (2011) Superparamagnetic iron oxide contrast agents: physicochemical characteristics and applications in MR imaging. *Eur. Radiol.*, **11**: 2319–31.
77. Sykova, E., and Jendelova, P. (2005) Magnetic resonance tracking of implanted adult and embryonic stem cells

in injured brain and spinal cord. *Ann. NY Acad. Sci.*, **1049**: 146–60.

78. Lee, H., Lee, E., Kim, D.K., Jang, N.K., Jeong, Y.Y., and Jon, S. (2006) Antibiofouling polymer-coated superparamagnetic iron oxide nanoparticles as potential magnetic resonance contrast agents for in vivo cancer imaging. *J. Am. Chem. Soc.*, **128**: 7383–9.
79. Beg, S., Rizwan, M., Sheick, A.M., Hasnain, M.S., Anwer, K., and Kohli, K. (2011) Advancement in carbon nanotubes: basics, biomedical applications and toxicity. *J. Pharm. Pharmacol.*, **63**: 141–63.
80. Li, Y., Cousins, B.G., Ulijn, R.V., and Kinloch, I.A. (2009) A study of the dynamic interaction of surfactants with graphite and carbon nanotubes using Fmoc-amino acids as a model system. *Langmuir*, **25**: 11760–7.
81. Antonelli, A., Serafini, S., Menotta, A., Sfara, C., Pierigè, F., et al. (2010) Improved cellular uptake of functionalized single-walled carbon nanotubes. *Nanotechnology*, **21**: 425101.
82. Salvador-Morales, C., Flahaut, E., Sim, E., Sloan, J., Green, M.L., and Sim, R.B. (2006) Complement activation and protein adsorption by carbon nanotubes. *Mol. Immunol.*, **43**: 193–201.
83. Liu, Z., Chen, K., Davis, C., Sherlock, S., Cao, Q., et al. (2008) Drug delivery with carbon nanotubes for in vivo cancer treatment. *Cancer Res.*, **68**: 6652–60.
84. Lay, C.L., Liu, Q.L., Tan, H.R., and Liu, Y. (2010) Delivery of paclitaxel by physically loading onto poly(ethylene glycol)(PEG)-graftcarbon nanotubes for potent cancer therapeutics. *Nanotechnology*, **21**: 065101.
85. Kam, N.W., O'Connell, M., Wisdom, J.A., and Dai, H. (2005) Carbon nanotubes as multifunctional

biological transporters and near-infrared agents for selective cancer cell destruction. *Proc. Natl. Acad. Sci. USA*, **102**: 11600–5.

86. Zhang, Z., Yang, X., Zhang, Y., Zeng, B., Wang, S., et al. (2006) Delivery of telomerase reverse transcriptase small interfering RNA in complex with positively charged single walled carbon nanotubes suppresses tumor growth. *Clin. Cancer Res.*, **12**: 4933–9.
87. Brennan, M.E., Coleman, J.N., Drury, A., Lahr, B., Kobayashi, T., and Blau, W.J. (2007) Nonlinear photoluminescence from van Hove singularities in multiwalled carbon nanotubes. *Optic Lett.*, **28**: 2654–65.
88. Park, E.J., Roh, J., Kim, S.N., Kang, M.S., Han, Y.A., et al. (2011) A single intratracheal instillation of single-walled carbon nanotubes induced early lung fibrosis and subchronic tissue damage in mice. *Arch. Toxicol.*, **85**: 1121–31.
89. Aillon, K.L., Xie, Y., El-Gendy, N., Berkland, C.J., and Forrest, M.L. (2009) Effects of nanomaterial physicochemical properties on in vivo toxicity. *Adv. Drug Deliv. Rev.*, **61**: 457–66.
90. Nasibulin, A.G., Pikhitsa, P.V., Jiang, H., Brown, D.P., Krasheninnikov, A.V., et al. (2007) A novel hybrid carbon material. *Nat. Nanotechnol.*, **2**: 156–61.
91. Yamakoshi, Y., Umezawa, N., Ryu, A., Arakane, K., Miyata, N., et al. (2003) Active oxygen species generated from photoexcited fullerene (C60) as potential medicines: O_2-*versus $1O_2$. *J. Am. Chem. Soc.*, **125**: 12803–9.
92. Iwamoto, Y., and Yamakoshi, Y. (2006) A highly water-soluble C60-NVP copolymer: a potential material for photodynamic therapy. *Chem. Commun.*, **46**: 4805–7.

93. Liu, J., Ohta, S., Sonoda, A., Yamada, M., Yamamoto, M., et al. (2007) Preparation of PEG-conjugated fullerene containing Gd3+ ions for photodynamic therapy. *J. Control Release*, **117**: 104–10.
94. Zakharian, T.Y., Seryshev, A., Sitharaman, B., Gilbert, B.E., Knight, V., and Wilson, L.J. (2005) A fullerene-paclitaxel chemotherapeutic: synthesis, characterization, and study of biological activity in tissue culture. *J. Am. Chem. Soc.*, **127**: 12508–9.
95. Abes, S., Ivanova, G.D., Abes, R., Arzumanov, A.A., Williams, D., et al. (2010) Bombesin functionalized gold nanoparticles show in vitro and in vivo cancer receptor specificity. *Proc. Natl. Acad. Sci. USA*, **107**: 8760–5.
96. Chanda, N., Kattamuri, V., Shukla, V., Zambre, A., Katti, K., Upendran, A., et al. (2010) Bombesin functionalized gold nanoparticles show in vitro and in vivo cancer receptor specificity. *Proc. Natl. Acad. Sci USA*, **107**: 8760–5.
97. Chen, X., Wang, X., Wang, Y., Yang, L., Hu, J., et al. (2010) Improved tumor-targeting drug delivery and therapeutic efficacy by cationic liposome modified with truncated bFGF peptide. *J. Control Release*, **145**: 17–25.
98. Desai, A., Vyas, T., and Amiji, M. (2008) Cytotoxicity and apoptosis enhancement in brain tumor cells upon coadministration of paclitaxel and ceramide in nanoemulsion formulations. *J. Pharm. Sci.*, **97**: 2745–56.
99. Pan, H., Soman, N.R., Schlesinger, P.H., Lanza, G.M., and Wickline, S.A. (2009) Cytolitic peptides nanoparticles ('NanoBees') for cancer therapy. *Wires Nanomed Nanobiotechnol.*, **3**: 318–27.
100. Choi, J., Burns, A.A., Williams, R.M., Zhou, Z., Flesken-Nikitin, A., et al. (2007) Core-shell silica

nanoparticles as fluorescent labels for nanomedicine. *J. Biomed. Opt.*, **12**: 064007.

101. He, X.X., Duan, J.H., Wang, K.M., Tan, W.H., Lin, X., and He, C.M. (2004) A novel fluorescent label based on organic dye-doped silica nanoparticles for HepG liver cancer cell recognition. *J. Nanosci. Nanotechnol.*, **4**: 585–9.
102. Santra, S., Liesenfeld, B., Dutta, D., Chatel, D., and Batich, C.D. (2005) Folate conjugated fluorescent silica nanoparticles for labeling neoplastic cells. *J. Nanosci. Nanotechnol.*, **5**: 899–904.
103. Santra, S., Yang, H., Dutta, D., Stanley, J.T., and Holloway, P.H. (2004) TAT conjugated FITC doped silica nanoparticles for bioimaging applications. *Chem. Commun.*, 2810–11.
104. Smith, J.E., Medley, C.D., Tang, Z.W., Shangguan, D.H., Lofton, C., and Tan, W.H. (2007) Aptamer-conjugated nanoparticles for the collection and detection of multiple cancer cells. *Anal. Chem.*, **79**: 3075–82.
105. Wang, L., Lofton, C., Popp, M., and Tan, W.H. (2007) Using luminescent nanoparticles as staining probes for Affymetrix GeneChips. *Bioconjugate Chem.*, **18**: 610–13.
106. Ravi Kumar, M.N.V., Sameti, M., Mohapatra, S.S., Kong, X., Lockey, R.F., et al. (2004) Cationic silica nanoparticles as gene carriers: Synthesis, characterization and transfection efficiency in vitro and in vivo. *J. Nanosci. Nanotechnol.*, **4**: 876–81.
107. Jin, S., and Ye, K.M. (2007) Nanoparticle-mediated drug delivery and gene therapy. *Biotechnol. Prog.*, **23**: 32–41.
108. Novoselov, K.S., Geim, A.K., Morozov, S.V., Jiang, D., Zhang, Y., et al. (2004) Electric field effect in atomically thin carbon films. *Science*, **306**: 666–9.

109. Zhang, K., Zhang, L.L., Zhao, X.S., and Wu, J. (2010) Graphene-polyaniline nanofiber composites as supercapacitor electrodes. *Chem. Mater.*, **22**: 1392–401.
110. Rao, C.N.R., Sood, A.K., Subrahmanyam, K.S., and Govindaraj, A. (2009) Graphene: the new two-dimensional nanomaterial. *Angew. Chem. Int. Ed.*, **48**: 7752–77.
111. Balandin, A.A., Ghosh, S., Bao, W.Z., Calizo, I., Teweldebrhan, D., et al. (2008) Superior thermal conductivity of single-layer graphene. *Nano Lett.*, **8**: 902–7.
112. Latil, S., and Henrard, L. (2006) Charge carriers in few-layer graphene films. *Phys. Rev. Lett.*, **97**: 036803–6.
113. Xuan, Y., Wu, Y.Q., Shen, T., Qi, M., and Capano, M.A. (2008) Atomic-layer-deposited nanostructures for graphene-based nanoelectronics. *Appl. Phys. Lett.*, **92**: 013101–3.
114. Yang, X.M., Tu, Y.F., Li, L., Shang, S., and Tao, X.M. (2010) Well-dispersed chitosan/graphene oxide nanocomposites. *ACS Appl. Mater. Interfaces*, **2**: 1707–1713.
115. Fan, H., Wang, L., Zhao, K., Li, N., Shi, Z., et al. (2010) Fabrication, mechanical properties, and biocompatibility of graphene-reinforced chitosan composites. *Biomacromolecules*, **11**: 2345–51.
116. Bai, H., Li, C., Wang, X., and Shi, G. (2010) A pH-sensitive graphene oxide composite hydrogel. *Chem. Commun.*, **46**: 2376–8.
117. Sun, S.T., and Wu, P.Y. (2011) A one-step strategy for thermal- and pH-responsive graphene oxide interpenetrating polymer hydrogel networks. *J. Mater. Chem.*, **21**: 4095–7.

118. Fang, F., Long, J., Zhao, W.F., Wang, L., and Chen, G. (2010) pH-Responsive chitosan-mediated graphene dispersions. *Langmuir*, **26**: 16771–4.
119. Liu, C., Alwarappan, S., Chen, Z.F., Kong, X., and Li, C.Z. (2010) Membraneless enzymatic biofuel cells based on graphene nanosheets. *Biosens. Bioelectron.*, **25**: 1829–33.
120. Stoller, M.D., Park, S., Zhu, Y.W., An, J., and Ruoff, R.S. (2008) Graphene-based ultracapacitors. *Nano Lett.*, **8**: 3498–502.
121. Wang, L., Lee, K., Sun, Y.Y., Lucking, M., Chen, Z., et al. (2009) Graphene oxide as an ideal substrate for hydrogen storage. *ACS Nano.*, **3**: 2995–3000.
122. Lu, C.H., Yang, H.H., Zhu, C.L., Chen, X., and Chen, G.N. (2009) A graphene platform for sensing biomolecules. *Angew. Chem. Int. Ed.*, **121**: 4879–81.
123. Qu, L.T., Liu, Y., Baek, J.B., and Dai, L. (2010) Nitrogen-doped graphene as efficient metal-free electrocatalyst for oxygen reduction in fuel cells. *ACS Nano*, **4**: 1321–6.
124. Feng, L.Z., and Liu, Z. (2011) Graphene in biomedicine: opportunities and challenges. *Nanomedicine*, **6**: 317–24.
125. Wang, Y., Li, Z.H., Wang, J., Li, J., and Lin, Y. (2011) Graphene and graphene oxide: biofunctionalization and applications in biotechnology. *Trends Biotechnol.*, **29**: 205–12.
126. Loh, K.P., Bao, Q.L., Eda, G., and Chhowalla, M. (2010) Graphene oxide as a chemically tunable platform for optical applications. *Nat. Chem.*, **2**: 1015–24.
127. Jiang, H.J. (2011) Chemical preparation of graphene-based nanomaterials and their applications in chemical and biological sensors. *Small*, **7**: 2413–27.

128. Guo, S.J., and Dong, S.J. (2011) Graphene nanosheet: synthesis, molecular engineering, thin film, hybrids, and energy and analytical applications. *Chem. Soc. Rev.*, **40**: 2644–72.
129. Shen, H., Zhang, L., Liu, M., and Zhang, Z. (2012) Biomedical applications of graphene. *Theranostics*, **2**: 283–94.
130. Zhang, Y., Nayak, T.R., Hong, H., and Cai, W. (2012) Graphene: a versatile nanoplatform for biomedical applications. *Nanoscale*, **4**: 3833–42.
131. Yang, K., Feng, L., Shi, X., and Liu, Z. (2013) Nano-graphene in biomedicine: theranostic applications. *Chem. Soc. Rev.*, **42**: 530–47.
132. Huang, X., Yin, Z., Wu, S., Qi, X., He, Q., et al. (2011) Graphene-based materials: synthesis, characterization, properties, and applications. *Small*, 7: 1876–902.
133. Liu, Z., Robinson, J.T., Sun, X., and Dai, H. (2008) PEGylated nanographene oxide for delivery of water-insoluble cancer drugs. *J. Am. Chem. Soc.*, **130**: 10876–7.
134. Bao, H., Pan, Y., Ping, Y., Sahoo, N.G., Wu, T., et al. (2011) Chitosan-functionalized graphene oxide as a nanocarrier for drug and gene delivery. *Small*, **7**: 1569–78.
135. Hu, H., Yu, J., Li, Y., Zhao, J., and Dong, H. (2012) Engineering of a novel pluronic F127/graphene nanohybrid for pH responsive drug delivery. *J. Biomed. Mater. Res. A*, **100**: 141–8.
136. Pan, Y.Z., Bao, H.Q., Sahoo, N.G., Wu, T.F., and Li, L. (2011) Water-soluble poly(N-isopropylacrylamide)–graphene sheets synthesized via click chemistry for drug delivery. *Adv. Funct. Mater.*, **21**: 2754–63.
137. Iyer, A.K., Khaled, G., Fang, J., and Maeda, H. (2006) Exploiting the enhanced permeability and retention

effect for tumor targeting. *Drug Discov. Today*, **11**: 812–18.

138. Sun, X., Liu, Z., Welsher, K., Robinson, J.T., Goodwin, A., et al. (2008) Nano-graphene oxide for cellular imaging and drug delivery. *Nano Res.*, **1**: 203–12.
139. Li, D., Müller, M.B., Gilje, S., Kaner, R.B., and Wallace, G.G. (2008) Processable aqueous dispersions of graphene nanosheets. *Nat. Nanotechnol.*, **3**: 101–5.
140. Li, X., Wang, X., Zhang, L., Lee, S., and Dai, H. (2008) Chemically derived, ultrasmooth graphene nanoribbon semiconductors. *Science*, **319**(5867): 1229–32.
141. Tasciotti, E., Liu, X., Bhavane, R., Plant, K., Leonard, A.D., et al. (2008) Mesoporous silicon particles as a multistage delivery system for imaging and therapeutic applications. *Nat. Nanotechnol.*, **3**: 151–7.
142. Lu, Y.J., Yang, H.W., Hung, S.C., Huang, C.Y., Li, S.M., et al. (2012) Improving thermal stability and efficacy of BCNU in treating glioma cells using PAA-functionalized graphene oxide. *Int. J. Nanomed.*, **7**: 1737–47.
143. Markovic, Z.M., Harhaji-Trajkovic, L.M., Todorovic-Markovic, B.M., Kepi , D.P., Arsikin, K.M., et al. (2011) In vitro comparison of the photothermal anticancer activity of graphene nanoparticles and carbon nanotubes. *Biomaterials*, **32**: 1121–9.
144. Wang, Y., Wang, K., Zhao, J., Liu, X., Bu, J., et al. (2013) Multifunctional mesoporous silica-coated graphene nanosheet used for chemo-photothermal synergistic targeted therapy of glioma. *J. Am. Chem. Soc.*, **135**: 4799–804.
145. Jaworski, S., Sawosz, E., Grodzik, M., Winnicka, A., Prasek, M., et al. (2013) In vitro evaluation of the

effects of graphene platelets on glioblastoma multiforme cells. *Int. J. Nanomed.*, **8**: 413–20.

146. Strakova, N., Ehrmann, J., Dzubak, P., Bouchal, J., and Kolar, Z. (2004) The synthetic ligand of peroxisome proliferator-activated receptor-γ ciglitazone affects human glioblastoma cell lines. *J. Pharmacol. Exp. Ther.*, **309**: 1239–47.
147. Stockmann-Juvala, H., Naarala, J., Loikkanen, J., Vähäkangas, K., and Savolainen, K. (2006) Fumonisin B1-induced apoptosis in neuroblastoma, glioblastoma and hypothalamic cell lines. *Toxicology*, **225**(2–3): 234–41.
148. Gao, X., Chung, L.W., and Nie, S. (2007) Quantum dots for in vivo molecules and cellular imaging. *Methods Mol. Biol.*, **374**: 135–45.
149. Tan, W.B., Jiang, S., and Zhang, Y. (2007) Quantum-dot based nanoparticles for targeted silencing of HER2/neu gene via RNA interference. *Biomaterials*, **28**: 1565–71.
150. Tekade, R.K., Kumar, P.V., Jain, N.K. (2009) Dendrimers in oncology: an expanding horizon. *Chem. Rev.*, **109**: 49–87.
151. Bharali, D.J., Khalil, M., Gurbuz, M., Simone, T.M., and Mousa, S.A. (2009) Nanoparticles and cancer therapy: a concise review with emphasis on dendrimers. *Int. J. Nanomed.*, **4**: 1–7.
152. Majoros, I.J., Myc, A., Thomas, T., Mehta, C.B., and Baker, J.R. Jr (2006) PAMAM dendrimer-based multifunctional conjugate for cancer therapy: synthesis, characterization, and functionality. *Biomacromolecules*, 7: 572–9.
153. Desai, P.N., Yuan, Q., and Yang, H. (2010) Synthesis and characterization of photocurable polyamidoamine dendrimer hydrogels as a versatile platform for tissue

engineering and drug delivery. *Biomacromolecules*, **11**: 666–73.

154. Ke, W.L., Shao, K., Huang, R.Q., Han, L., Liu, Y., et al. (2009) Gene delivery targeted to brain using an Angiopep-conjugated polyethyleneglycol-modified polyamidoamine dendrimer. *Biomaterials*, **30**: 6976–85.
155. Agrawal, A., Min, D.H., Singh, N., Zhu, H., Birjiniuk, A., et al. (2009) Functional delivery of siRNA in mice using dendriworms. *ACS Nano*, **3**: 2495–504.
156. Yang, W., Barth, R.F., Wu, G., Huo, T., Tjarks, W., et al. (2009) Convection enhanced delivery of boronated EGF as a molecular targeting agent for neutron capture therapy of brain tumors. *J. Neurooncol.*, **95**: 355–65.
157. Zhang, L., Zhu, S., Qian, L., Pei, Y., Qiu, Y., and Jiang, Y. (2011) RGD-modified PEG-PANAM-DOX conjugates: in vitro and in vivo studies for glioma. *Eur. J. Pharm. Biopharm.*, **79**: 232–40.
158. Klajnert, B., and Bryszewska, M. (2001) Dendrimers: properties and applications. *Acta Biochim. Pol.*, **48**: 199–208.
159. Bernardi, R.J., Lowery, A.R., Thompson, P.A., Blaney, S.M., and West, J.L. (2008) Immunonanoshells for targeted photothermal ablation in medulloblastoma and glioma: an in vitro evaluation using human cell lines. *J. Neurooncol.*, **86**: 165–72.
160. Koo, Y.L., Reddy, G.R., Bhojani, M., Schneider, R., Philbert, M.A., et al. (2006) Brain cancer diagnosis and therapy with nanoplatforms. *Adv. Drug Deliv. Rev.*, **58**: 1556–7.
161. Sun, C., Veiseh, O., Gunn, J., Fang, C., Hansen, S., et al. (2008) In vivo MRI detection of gliomas by chlorotoxin-conjugated superparamagnetic nanoprobes. *Small*, **4**: 372–9.

162. Wang, C., Huang, Y., and Peng, C. (2009) Photothermal ablation of stem-cell like glioblastoma using carbon nanotubes functionalized with anti-CD133. *ICBME Proc.*, **23**: 888–91.

163. Madhankumar, A.B., Slage-Webb, B., Mintz, A., Sheehan, J.M., and Connor, J.R. (2006) Interleukin-13 receptor-targeted nanovesicles are a potential therapy for glioblastoma multiforme. *Mol. Cancer Ther.*, **5**: 3162–9.

164. Lu, W., Sun, Q., Wan, J., She, Z., and Jiang, X. (2006) Cationic albumin-conjugated pegylated Nanoparticles allow gene delivery into brain tumors via intravenous administration. *Cancer Res.*, **66**: 11878–86.

165. Reddy, G.R., Bhojani, M.S., McConville, P., Moody, J., Moffat, B.A., et al. (2006) Vascular targeted Nanoparticles for imaging and treatment of brain tumors. *Clin. Cancer Res.*, **12**: 6677–86.

166. Bernardi, A., Braganhol, E., Jager, E., Figueiró, F., Edelweiss, M.I., et al. (2009) Indomethacin-loaded nanocapsules treatment reduces in vivo glioblastoma growth in a rat glioma model. *Cancer Lett*, **281**: 53–63.

167. Wang, X., Pang, Y., Ku, G., Xie, X., Stoica, G., and Wang, L.W. (2003) Noninvasive laser-induced photoacoustic tomography for structural and functional in vivo imaging of the brain. *Nat. Biotechnol.*, **21**: 803–6.

168. Pan, D., Pramanik, M., Senpan, A., Allen, J.S., Zhang, H., et al. (2011) Molecular photoacoustic imaging of angiogenesis with integrin-targeted gold nanobeacons. *Faseb J.*, **25**: 875–82.

169. Li, M., Oh, J., Xie, X., Ku, G., Wei, W., et al. (2008) Simultaneous molecular and hypoxia imaging of brain tumors in vivo using spectroscopic photoacoustic tomography. *Proceedings IEEE*, **96**: 481–9.

170. Chertok, B., David, A.E., Huang, Y., and Yang, V.C. (2007) Glioma selectivity of magnetically targeted Nanoparticles: a role of abnormal tumor hydrodynamics. *J. Control Release*, **122**: 315–23.
171. Schneider, T., Becker, A., Ringe, K., Reinhold, A., Firsching, R., and Sabel, B.A. (2008) Brain tumor therapy by combined vaccination and antisense oligonucleotide delivery with nanoparticles. *J. Neuroimmunol.*, **195**: 21–7.
172. Wu, Z.M., Yuan, X.H., Jiang, P.C., Li, Z.Q., and Wu, T. (2006) Antisense oligonucleodes targeting the focal adhesion kinase inhibit proliferation, induce apoptosis and cooperate with cytotoxic drugs in human glioma cells. *J. NeuroOncol.*, **77**: 117–23.
173. Chang, S.M., Kuhn, J.G., Robins, H.I., Schold, S.C. Jr, Spence, A.M., et al. (2001) A phase II study of paclitaxel in patients with recurrent malignant glioma using different doses depending upon the concomitant use of anticonvulsants: a North American brain tumor consortium report. *Cancer*, **91**: 417–22.
174. Xin, H., Chen, L., Gu, J., Ren, X., Wei, Z., et al. (2010) Enhanced anti-glioblastoma efficacy by PTX-loaded PEGylated poly(ε-caprolactone) nanoparticles: in vitro and in vivo evaluation. *Int. J. Pharmacol.*, **402**: 238–47.
175. Xin, H., Jiang, X., Gu, J., Sha, X., Chen, L., et al. (2011) Angiopep-conjugated poly(ethylene glycol)-co-poly(e-caprolactone) nanoparticles as dual-targeting drug delivery system for brain glioma. *Biomaterials*, **32**: 4293–305.
176. Chien, A.J., Illi, J.A., Ko, A.H., Korn, W.H., Fong, L., et al. (2009) A phase I study of a 2-day lapatinib chemosensitization pulse preceding Nanoparticle

albumin-bound Paclitaxlel for advanced solid malignancies. *Clin. Cancer Res.*, **15**: 5569–75.

177. Majoros, I.J., Myc, A., Thomas, T., Mehta, C.B., and Baker, J.R. Jr (2006) PAMAM dendrimer-based multifunctional conjugate for cancer therapy: synthesis, characterization, and functionality. *Biomacromolecules*, 7: 572–9.
178. Caruso, G., Caffo, M., Raudino, G., Alafaci, C., Salpietro, F.M., and Tomasello, F. (2010) Antisense oligonucleotides as an innovative therapeutic strategy in the treatment of high-grade gliomas. *Rec. Pat. CNS Drug Discov.*, **5**: 53–69.
179. Caruso, G., Caffo, M., Raudino, G., Alafaci, C., and Tomasello, F. (2011) New therapeutic strategies in gliomas treatment. In: Abujamra, A.L. (ed.) *Brain Tumors – Current and Emerging Therapeutic Strategies*, pp. 281–306. Rijeka: InTech.
180. Caruso, G., Caffo, M., Alafaci, C., Raudino, G., Cafarella, D., et al. (2011) Could nanoparticles systems have a role in the treatment of cerebral gliomas? *Nanomedicine*, 7: 744–52.
181. Caruso, G., Raudino, G., Caffo, M., Alafaci, C., and Granata, F. (2010) Nanotechnology platforms in diagnosis and treatments of primary brain tumors. *Recent Pat. Nanotechnol.*, **4**: 119–22.
182. Nair, B.G., Varghese, S.H., Nair, R., Yoshida, Y., Maekawa, T., and Kumar, D.S. (2011) Nanotechnology platforms; an innovative approach to brain tumor therapy. *Med. Chem.*, 7: 488–503.
183. Collins, K.L., Orringer, D.A., and Patil, P.G. (2010) Nanotechnology in neurosurgery. *J. Nanotechnol. Eng. Med.*, **1**: 1–15.
184. Caffo, M., Caruso, G., Barresi, V., Pino, M.A., Venza, M., and Tomasello, F. (2012) Immunohistochemical

study of CD68 and CR3/43 in astrocytic gliomas. *J. Anal. Oncol.*, **1**: 42–9.

185. Raudino, G., Caffo, M., Caruso, G., Alafaci, C., and Tomasello, F. (2011) Nanoparticle-based cerebral drug delivery systems and antiangiogenic approach in gliomas treatment. *Rec. Pat. Nanotech.*, **5**: 239–44.
186. Wang, Y., and Luang, L. (2012) Multifunctional theranostic nanoparticles for brain tumors. *Mol. Ther.*, **20**: 1.
187. Im, S.A., Gomez-Manzano, C., Fueyo, J., Liu, T.J., Ke, L.D., et al. (1999) Antiangiogenesis treatment for gliomas: transfer of antisense-vascular endothelial growth factor inhibits tumor growth in vivo. *Cancer Res.*, **59**: 895–900.
188. Agemy, L., Friedmann-Morvinski, D., Kotamraju, V.R., Roth, L., Sugahara, K.N., et al. (2011) Targeted nanoparticle enhanced proapoptotic peptide as potential therapy for glioblastoma. *Proc. Natl. Acad. Sci. USA*, **42**: 17450–5.
189. Sawyer, A.J., Saucier-Sawyer, J.K., Booth, C.J., Liu, J., Patel, T., et al. (2011) Convection-enhanced delivery of camptothecin-loaded polymer nanoparticles for treatment of intracranial tumors. *Drug Deliv. Transl. Res.*, **1**: 34–42.
190. Hadjipanayis, C.G., Machaidze, R., Kaluzova, M., Wang, L., Schuette, A.J., et al. (2010) EGFRvIII antibody conjugated iron oxide nanoparticles for MRI guided convection-enhanced delivery and targeted therapy of glioblastoma. *Cancer Res.*, **70**: 6303–12.
191. Kim, T., Jeong, Y., Jin, S., Pei, J., Jung, T.Y., et al. (2011) Preparation of polylactide-co-glycolide nanoparticles incorporating celecoxib and their antitumor activity against brain tumor cells. *Int. J. Nanomed.*, **6**: 2621–31.

192. Lim, K.J., Bisht, S., Bar, E.E., Maitra, A., and Eberhart, C.G. (2011) A polymeric nanoparticle formulation of curcumin inhibits growth, clonogenicity and stem-like fraction in malignant brain tumors. *Cancer Biol. Ther.*, **11**: 464–73.
193. Jung, J., Solanki, A., Memoli, K.A., Kamei, K., Kim, H., et al. (2010) Selective inhibition of human brain tumor cell proliferation via multifunctional quantum dot-based siRNA delivery. *Angew. Chem. Int. Ed. Engl.*, **49**: 103–7.
194. Zhao, D., Alizadeh, D., Zhang, L., Liu, W., Farrukh, O., et al. (2011) Carbon nanotubes enhance CpG uptake and potentiate antiglioma immunity. *Clin. Cancer Res.*, **17**: 771–82.
195. Alizadeh, D., Zhang, L., Hwang, J., Schluep, T., and Badie, B. (2010) Tumor-associated macrophages are predominant carriers of cyclodextrin-based nanoparticles into gliomas. *Nanomedicine*, **6**: 382–90.
196. Rozhkova, E.A., Ulasov, I., Lai, B., Dimitrijevic, N.M., Lesniak, M.S., and Rajh, T. (2009) A high-performance nanobio phocatalyst for targeted cancer-cell therapy. *Nano Lett.*, **9**: 3337–42.
197. Rajh, T., Dimitrijevic, N.M., and Rozhkova, E.A. (2011) Titanium dioxide nanoparticles in advanced imaging and nanotherapeutics. *Methods Mol. Biol.*, **726**: 63–75.
198. Steiniger, S.C.J., Kreuter, J., Khalansky, A.S., Skidan, I.N., Bobruskin, A.I., et al. (2004) Chemotherapy of glioblastoma in rats using doxorubicin-loaded nanoparticles. *Int. J. Cancer* **109**: 759–67.
199. Orringer, D.A., Koo, Y., Kopelman, R., and Sagher, O. (2009) Small solutions for big problems: the application of nanoparticles to brain tumor diagnosis and therapy. *Clin. Pharmacol. Ther.*, **85**: 531–4.

200. Zara, G.P., Cavalli, R., Bargoni, A., Fundarò, A., and Vighetto, D. (2002) Intravenous administration to rabbits of non-stealth and stealth doxorubicin-loaded solid lipid nanoparticles at increasing concentrations of stealth agent: pharmacokinetics and distribution of doxorubicin in brain and other tissues. *J. Drug Target*, **10**: 327–35.
201. Kuroda, J., Kuratsu, J., and Yasunaga, M. (2009) Potent antitumor effect of SN-38-incorporating polymeric micelle, NK012, against malignant glioma. *Int. J. Cancer*, **124**: 2505–11.
202. Ljubimova, J.Y., Fujita, M., Khazenzon, N.M., Lee, B.S., Wachsmann-Hogiu, S., et al. (2008) Nanoconjugate based on polymalic acid for tumor targeting. *Chem. Biol. Interact.*, **171**: 195–203.
203. Chertok, B., David, A.E., and Yang, V.C. (2010) Polyethyleneimine-modified iron oxide nanoparticles for brain tumor drug delivery using magnetic targeting and intra-carotid administration. *Biomaterials*, **31**: 6317–24.
204. Etame, A.B., Diaz, R.J., O'Reilly, M.A., Smith, C.A., Mainprize, T.G., et al. (2012) Enhanced delivery of gold nanoparticles with therapeutical potential into the brain using MRI-guided focused ultrasound. *Nanomedicine*, **8**: 1133–42.
205. Caruso, G., Caffo, M., Raudino, G., and Tomasello, F. (2012) Nanomedicine and nanotechnology. In Al-Jumaily A (ed.), Nanoparticles and brain tumor treatment, pp. 14–32. New York: ASME Press.
206. Maier-Hauff, K., Ulrich, F., Nestler, D., Niehoff, H., Wust, P., et al. (2011) Efficacy and safety of intratumoral thermotherapy using magnetic iron-oxide nanoparticles combined with external beam radiotherapy on patients with recurrent glioblastoma multiforme. *J. Neurooncol.*, **103**: 317–24.

207. Chang, W.H., Kao, C.-H., Lin, C.-I., and Wang, S.-J.: US0154397 (2007).
208. Vinod, D.L., Tapan, K.J., and Diandra, L.-P.: US0264199 (2007).
209. Akhtari, M., and Engel, J.: US0206146 (2008).
210. Caruso, G., Raudino, G., and Caffo, M. (2013) Patented nanomedicines for the treatment of brain tumors. *Pharm. Pat. Analyst*, **2**: 745–54.
211. Akhtari, M., and Engel, J.: US0044911 (2011).
212. Yang, V.C., Chertok, B., and David, A.E.: US0054236 (2011).
213. Chithrani, B.D., Ghazani, A.A., and Chan, W.C. (2006) Determining the size and shape dependence of gold nanoparticle uptake into mammalian cells. *Nano Lett.*, **6**: 662–8.
214. Bayford, R.H., Demosthenous, A., Iles, R.K., Rademacher, T.W., and Roitt, I.M.: US0212029 (2011).
215. Hainfeld, J.F., and Slatkin, D.N.: US7530940 (2009).
216. Caruso, G., Caffo, M., Raudino, G., Tomasello, C., Alafaci, C., and Tomasello, F. (2012) Nanomedicine and brain tumors treatment. In: Souto, E.B. (ed.) *Patenting Nanomedicines. Legal Aspects, Intellectual Property and Grant Opportunities*, pp. 167–204. Berlin, Heidelberg: Springer-Verlag.
217. Choonara, Y.E.: EP-2370055-A (2011).
218. Black, K.L.: EP-2509421-A (2012).
219. Johanson, L.M., Lee, M.S., Nichols, M.J., and Zimmerman, G.R.: EP-1883407-A (2005).
220. Drummond, D., and Kirpotin, D.: US0110798 (2007).
221. Hongbin, H.: US0239000 (2012).
222. Wu, X.Y., and Shalviri, A.: WO 2013127004 A1 (2013).
223. Chavanpatil Mahesh, D., and Panyam Jayanth: WO 2008024753 A2 (2008).

224. Alkon, L.D., Nelson, J.T., and Quattrone, A.: EP-1581186-A (2005).
225. Forte, T.M., and Nikanjam, M.: US8252338 (2010).
226. Vitol, S., Peterson, C., Larsson, O., Holm, P., and Aberg, B. (1992) Elevated uptake of low density lipoproteins by human lung cancer tissue in vivo. *Cancer Res.*, **52**: 6244–7.
227. Lum, D.F., McQuaid, K.R., Gilbertson, V.L., and Hughes-Fulford, M. (1999) Coordinate up-regulation of low-density lipoprotein receptor and cyclo-oxygenase-2 gene expression in human colorectal cells and in colorectal adenocarcinoma biopsies. *Int. J. Cancer* **83**: 162–6.
228. Jung-Testas, I., Weintraub, H., Dupuis, D., Eychenne, B., Baulieu, E.E., and Robel, P. (1992) Low density lipoprotein-receptors in primary cultures of rat glial cells. *J. Steroid Biochem. Mol. Biol.*, **42**: 597–605.
229. Heckl, S., Braun, K., Pipkorn, R., and Waldeck, R.: US0123699 (2007).
230. Ferrari, M., Tasciotti, E., and Quattrocchi, N.: EP-2547329-A (2013).
231. Matsudai, M., and Hunt, G. (2005) Nanotechnology and public health. *Nippon Koshu Eisei Zasshi*, **52**: 923–7.
232. Hoet, P.H., Bruske-Hohlfeld, I., and Salata, O.V. (2004) Nanoparticles – known and unknown health risks. *J Nanobiotechnol.*, **2**: 12.
233. Williams, D. (2004) Nanotechnology: a new look. *Med. Device Technol.*, **15**: 9–10.
234. Williams, D. (2005) The risks of nanotechnology. *Med. Device Technol.*, **16**: 6, 9–10.

5

Future scenarios: nanoparticles and stem cells

DOI: 10.1533/9781908818744.151

Abstract: Although nanotechnologies hold great promise in glioma therapy, there have been few positive results in the clinical setting so far due to intra- and inter-patient heterogeneity. New models allowing tumor-specific targeting and extensive and safe intratumoral distribution must be developed to efficiently deliver drugs carried by nanoparticles. Laboratory modifications of nanoparticles have been tested to specifically track tumor cells, and a great step forward has been made thanks to the conjugation with neural and mesenchymal stem cells. These stem cells possess a specific tropism for brain tumors which makes them putative candidates as delivery vehicles for nanoparticles in glioma therapy. Stem cells have a natural tendency to migrate and distribute within the tumor mass and they can also incorporate nanoparticles. Stem cell therapy combined with nanotechnology could be a promising tool to efficiently deliver drugs to brain tumors.

Key words: brain tumor, drug delivery, nanoparticles, nanotechnology, neural stem cells.

Published by Elsevier Limited, 2014

Adult neural stem cells (NSCs) are self-renewing, proliferative, multipotent cells that give rise to the different neuroectodermal-derived cell populations, including neurons and glia. In the adult mammalian brain, these cells can principally be isolated from the subgranular zone of the dentate gyrus and the anterior area of the subventricular zone along the ventricle.[1] These sources of autologous NSCs are scarce, so for clinical applications fetal brain, adult allogeneic brain, and embryonic stem cells are used as substitutes.[2] NSCs possess extensive tropism for experimental intracranial gliomas in adult rodents.[3,4] Strong antineoplastic effects were reported following intracranial administration of gene-modified NSCs. The molecules tested ranged from cytokines,[5–7] to tumor necrosis factor-related apoptosis-inducing ligand,[8–10] or toxic molecules,[11–16] the fragment of human metalloproteinase 2, PEX,[17] antiangiogenic agents,[18,19] and finally cytokines combined with toxic molecules.[16] NSCs have also been used as cell carriers for replication, transportation, and local release of oncolytic adenoviruses into tumors.[20] A pilot clinical trial to evaluate the safety of genetically modified embryonic NSCs expressing the suicide gene (*Escherichia coli* cytosine deaminase) combined with oral administration of 5-fluorocytosine for the treatment of recurrent high-grade gliomas is actually ongoing (*http://clinicaltrials.gov/show/NCT01172964*). However, clinical application of NSCs, although appealing, is still limited by ethical problems associated with their isolation and by potential immunological incompatibility due to the requirement for allogeneic transplantation.

Fewer ethical problems are coupled to the use of mesenchymal stem cells (MSCs). These cells can also be easily propagated in vitro. MSCs are usually isolated from bone marrow,[21,22] anyway similar cells are reported as being present in adipose tissue,[23] umbilical cord blood,[24–27] peripheral blood,[28,29] connective tissues of the dermis, and skeletal muscle.[30] MSCs are cultured under specific conditions: they have to be induced by a panel of nonspecific surface antigens (CD29, CD44, CD90, CD73) to give rise to different cell lines: adipocytes, chondrocytes, and osteoblasts. Different subpopulations of MSCs have been identified under particular culture conditions: in low oxygen tension, for example, it has been possible to isolate the so-called Marrow-Isolated Adult Multilineage Inducible (MIAMI) MSCs, which are phenotypically unique and express embryonic stem cell markers. These cells, without distinction for the age of human donors, can differentiate into cell lines derived from all three germ layers. So they can develop into neuron-like cells,[31,32] thus representing autologous human cell populations for tissue regeneration of skeletal and nervous system disorders. This multistep process passes through their differentiation into NSCs, neural/ neuronal progenitors, and finally neuron-like cells with electrophysiological characteristics that are similar to those observed in neurons. From their last differentiation onwards, MSCs can assimilate themselves with NSCs, without large differences. As for NSCs, MSCs possess the capacity to migrate to areas of injury and tumors. This characteristic makes them pivotal in therapeutics. In fact, after intratumoral or contralateral injection of MSCs into mouse models of glioma, MSCs are found all around the tumor, at the border between the tumor mass and the brain parenchyma. Furthermore, MSCs follow tumor infiltrations, migrating and localizing around glioma cells. The pathways involved can be different and rely on numerous factors, the most studied of which are stromal-derived factor-1,[33] platelet-derived growth factor,[34] epidermal growth factor,[35] matrix metalloproteinase-1,[36] and macrophage chemoattractant protein-1.[37] As described for

NSCs, MSCs that have been genetically modified with viruses expressing immunostimulators or toxic molecules have been shown to impede tumor growth in animal models.[38–49] MSCs have also been used to deliver intact oncolytic adenoviruses into tumors.[50,51] Due to their ease of isolation, MSCs are a promising tool as therapeutic gene-delivery vectors. However, no clinical trials have been performed, probably because the viral carriers used for the therapeutic gene delivery carry risks of toxicity, immunogenicity, and insertional mutagenesis, and have high manufacturing costs.[52,53]

Thanks to their unique characteristics, NSCs and MSCs are promising in brain tumor treatment, alone or in combination with other new technologies. Nanotechnology can aid in monitoring the presence of stem cells and stem cells can be tracked if nanoparticles (NPs) are incorporated into them, such as superparamagnetic iron oxide NPs, fluorochrome-loaded NPs, or quantum dots.[54–59] The migratory behavior of stem cells towards glioma cells in vivo has been routinely demonstrated with this approach.[60–64] NP systems have also been used for gene transfection. For example, biodegradable polymer NPs have been developed to deliver the vascular endothelial growth factor gene to human MSCs and human embryonic stem cell-derived cells to promote angiogenesis.[65] Another application of NPs with stem cells is intracellular delivery of growth factors to induce osteogenic or chondrogenic differentiation of stem cells.[66–72] Stem cells were also transfected using NPs encapsulated with plasmid DNA encoding bone morphogenetic protein-2 to induce their odontogenic or osteogenic differentiation.[73,74] The combination of stem cells and drug-loaded NPs for therapeutic applications in glioma therapy is a promising strategy. Roger et al.[75] combined MIAMI cells as carriers with a subpopulation of MSCs with polylactide-NPs and lipid nanocapsules to target brain tumors and showed that

MIAMI cells mainly localized at the border between tumor cells and normal brain parenchyma following intratumoral or contralateral injection into mice with U87MG glioma. Polylactide-NPs and lipid nanocapsules were efficiently internalized into MIAMI cells whereas cell viability and differentiation were not affected. Furthermore, these NP-loaded cells were able to migrate towards the U87MG experimental human glioma model.[75] The therapeutic efficacy of MIAMI cells carrying drug-loaded NPs was evaluated by using the cytotoxic compound 1,1-di(4-hydroxyphenyl)-2-ferrocenylbut-1ene (Fc-diOH), an analog of 4-hydroxy-tamoxifen. MIAMI cells loaded with Fc-diOH–lipid nanocapsules were toxic to the U87MG glioma cell line both in vitro and in vivo. This indicates that MIAMI cells are able to internalize drug-loaded NPs and deliver the cytotoxic agent into the tumor. Agents other than drugs could be encapsulated into NPs, such as DNA to transfer genes into the tumor or miRNA to repress the translation of mRNAs.

The use of stem cells, in particular MSCs, as cellular carriers is a promising therapeutic strategy to deliver specific drug-loaded NPs. However, the interaction between MSCs and tumor cells should be determined. Currently, several studies highlight that caution should be used in the therapeutic exploitation of MSCs for malignant conditions. Indeed, the exact biological function of MSCs in brain tumors and peripheral tumors is still unclear.[76] Some studies indicate that MSCs promote tumor development either by providing a niche for cancer stem cells, through impairing immune surveillance or by differentiation into cancer-associated fibroblast-like cells.[77–79] Other studies have demonstrated an opposite effect[38,80] or no effect at all.[45,81–83] Before using MSCs as cellular carriers, the fate of these cells within the brain tumor still needs to be carefully evaluated.

References

1. Ming, G.L., and Song, H. (2005) Adult neurogenesis in the mammalian central nervous system. *Annu. Rev. Neurosci.*, **28**: 223–50.
2. Singh, G. (2001) Sources of neuronal material for implantation. *Neuropathology*, **21**: 110–14.
3. Aboody, K.S., Brown, A., Rainov, N.G., Bower, K.A., Liu, S., et al. (2000) Neural stem cells display extensive tropism for pathology in adult brain: evidence from intracranial gliomas. *Proc. Natl. Acad. Sci. USA*, **97**: 12846–51.
4. Benedetti, S., Pirola, B., Pollo, B., Magrassi, L., Bruzzone, M.G., et al. (2000) Gene therapy of experimental brain tumors using neural progenitor cells. *Nat. Med.*, **6**: 447–50.
5. Ehtesham, M., Kabos, P., Kabosova, A., Neuman, T., Black, K.L., and Yu, J.S. (2002) The use of interleukin 12-secreting neural stem cells for the treatment of intracranial glioma. *Cancer Res.*, **62**: 5657–63.
6. Yang, S.Y., Liu, H., and Zhang, J.N. (2004) Gene therapy of rat malignant gliomas using neural stem cells expressing IL-12. *DNA Cell Biol.*, **23**: 381–9.
7. Yuan, X., Hu, J., Belladonna, M.L., Black, K.L., and Yu, J.S. (2006) Interleukin-23-expressing bone marrow-derived neural stem-like cells exhibit antitumor activity against intracranial glioma. *Cancer Res.*, **66**: 2630–8.
8. Shah, K., Tung, C.H., Yang, K., Weissleder, R., and Breakefield, X.O. (2004) Inducible release of TRAIL fusion proteins from a proapoptotic form for tumor therapy. *Cancer Res.*, **64**: 3236–42.
9. Ehtesham, M., Kabos, P., Gutierrez, M.A., Chung, N.H., Griffith, T.S., et al. (2002) Induction of glioblastoma apoptosis using neural stem cell-mediated

delivery of tumor necrosis factor-related apoptosis-inducing ligand. *Cancer Res.*, **62**: 7170–4.

10. Corsten, M.F., Miranda, R., Kasmieh, R., Krichevsky, A.M., Weissleder, R., and Shah, K. (2007) MicroRNA-21 knockdown disrupts glioma growth in vivo and displays synergistic cytotoxicity with neural precursor cell delivered S-TRAIL in human gliomas. *Cancer Res.*, **67**: 8994–9000.
11. Barresi, V., Belluardo, N., Sipione, S., Mudo, G., Cattaneo, E., and Condorelli, D.F. (2003) Transplantation of prodrug-converting neural progenitor cells for brain tumor therapy. *Cancer Gene Ther.*, **10**: 396–402.
12. Li, S., Tokuyama, T., Yamamoto, J., Koide, M., Yokota, N., and Namba, H. (2005) Potent bystander effect in suicide gene therapy using neural stem cells transduced with herpes simplex virus thymidine kinase gene. *Oncology*, **69**: 503–8.
13. Li, S., Tokuyama, T., Yamamoto, J., Koide, M., Yokota, N., and Namba, H. (2005) Bystander effect-mediated gene therapy of gliomas using genetically engineered neural stem cells. *Cancer Gene Ther.*, **12**: 600–7.
14. Li, S., Gao, Y., Tokuyama, T., Yamamoto, J., Yokota, N., et al. (2007) Genetically engineered neural stem cells migrate and suppress glioma cell growth at distant intracranial sites. *Cancer Lett.*, **251**: 220–7.
15. Zhao, Y., and Wang, S. (2010) Human NT2 neural precursor-derived tumor-infiltrating cells as delivery vehicles for treatment of glioblastoma. *Hum. Gene Ther.*, **21**: 683–94.
16. Ito, S., Natsume, A., Shimato, S., Ohno, M., Kato, T., et al. (2010) Human neural stem cells transduced with IFN-β and cytosine deaminase genes intensify bystander effect in experimental glioma. *Cancer Gene Ther.*, **17**: 299–306.

17. Kim, S.K., Cargioli, T.G., Machluf, M., Yang, W., Sun, Y., et al. (2005) PEX-producing human neural stem cells inhibit tumor growth in a mouse glioma model. *Clin. Cancer Res.*, **11**: 5965–70.
18. Lorico, A., Mercapide, J., Solodushko, V., Alexeyev, M., Fodstad, O., and Rappa, G. (2008) Primary neural stem/progenitor cells expressing endostatin or cytochrome P450 for gene therapy of glioblastoma. *Cancer Gene Ther.*, **15**: 605–15.
19. van Eekelen, M., Sasportas, L.S., Kasmieh, R., Yip, S., Figueiredo, J.L., et al. (2010) Human stem cells expressing novel TSP-1 variant have anti-angiogenic effect on brain tumors. *Oncogene*, **29**: 3185–95.
20. Tyler, M.A., Ulasov, I.V., Sonabend, A.M., Nandi, S., Han, Y., et al. (2009) Neural stem cells target intracranial glioma to deliver an oncolytic adenovirus in vivo. *Gene Ther.*, **16**: 262–78.
21. Friedenstein, A.J., Piatetzky, II S., and Petrakova, K.V. (1966) Osteogenesis in transplants of bone marrow cells. *J. Embryol. Exp. Morphol.*, **16**: 381–90.
22. Pittenger, M.F., Mackay, A.M., Beck, S.C., Jaiswal, R.K., Douglas, R., et al. (1999) Multilineage potential of adult human mesenchymal stem cells. *Science*, **284**(5411): 143–7.
23. Zuk, P.A., Zhu, M., Mizuno, H., Huang, J., Futrell, J.W., et al. (2001) Multilineage cells from human adipose tissue: implications for cell-based therapies. *Tissue Eng.*, 7: 211–28.
24. Bieback, K., Kern, S., Kluter, H., and Eichler, H. (2004) Critical parameters for the isolation of mesenchymal stem cells from umbilical cord blood. *Stem Cells*, **22**: 625–34.
25. Erices, A., Conget, P., and Minguell, J.J. (2000) Mesenchymal progenitor cells in human umbilical cord blood. *Br. J. Haematol.*, **109**: 235–42.

26. Goodwin, H.S., Bicknese, A.R., Chien, S.N., Bogucki, B.D., Quinn, C.O., and Wall, D.A. (2001) Multilineage differentiation activity by cells isolated from umbilical cord blood: expression of bone, fat, and neural markers. *Biol. Blood Marrow Transplant*, **7**: 581–8.
27. Kogler, G., Sensken, S., Airey, J.A., Trapp, T., Muschen, M., et al. (2004) A new human somatic stem cell from placental cord blood with intrinsic pluripotent differentiation potential. *J. Exp. Med.*, **200**: 123–35.
28. Zvaifler, N.J., Marinova-Mutafchieva, L., Adams, G., Edwards, C.J., Moss, J., et al. (2000) Mesenchymal precursor cells in the blood of normal individuals. *Arthritis Res.*, **2**: 477–88.
29. Kuznetsov, S.A., Mankani, M.H., Gronthos, S., Satomura, K., Bianco, P., and Robey, P.G. (2001) Circulating skeletal stem cells. *J. Cell Biol.*, **153**: 1133–40.
30. Jiang, Y., Vaessen, B., Lenvik, T., Blackstad, M., Reyes, M., and Verfaillie, C.M. (2002) Multipotent progenitor cells can be isolated from postnatal murine bone marrow, muscle, and brain. *Exp. Hematol.*, **30**: 896–904.
31. D'Ippolito, G., Howard, G.A., Roos, B.A., and Schiller, P.C. (2006) Isolation and characterization of marrow-isolated adult multilineage inducible (MIAMI) cells. *Exp. Hematol.*, **34**: 1608–10.
32. Tatard, V.M., D'Ippolito, G., Diabira, S., Valeyev, A., Hackman, J., et al. (2007) Neurotrophin-directed differentiation of human adult marrow stromal cells to dopaminergic-like neurons. *Bone*, **40**: 360–73.
33. Wynn, R.F., Hart, C.A., Corradi-Perini, C., O'Neill, L., Evans, C.A., et al. (2004) A small proportion of mesenchymal stem cells strongly expresses functionally active CXCR4 receptor capable of promoting migration to bone marrow. *Blood*, **104**: 2643–5.

34. Fiedler, J., Roderer, G., Gunther, K.P., and Brenner, R.E. (2002) BMP-2, BMP-4, and PDGF-bb stimulate chemotactic migration of primary human mesenchymal progenitor cells. *J. Cell Biochem.*, **87**: 305–12.
35. Sato, H., Kuwashima, N., Sakaida, T., Hatano, M., Dusak, J.E., et al. (2005) Epidermal growth factor receptor-transfected bone marrow stromalcells exhibit enhanced migratory response and therapeutic potential againstmurine brain tumors. *Cancer Gene Ther.*, **12**: 757–68.
36. Ho, I.A., Chan, K.Y., Ng, W.H., Guo, C.M., Hui, K.M., et al. (2009) Matrix metalloproteinase 1 is necessary for the migration of human bone marrowderived mesenchymal stem cells toward human glioma. *Stem Cells*, **27**: 1366–75.
37. Xu, J., Futakuchi, M., Iigo, M., Fukamachi, K., Alexander, D.B., et al. (2010) Involvement of macrophage inflammatory protein 1α (MIP1α) in promotion of rat lung and mammary carcinogenic activity of nanoscale titanium dioxide particles administered by intra-pulmonary spraying. *Carcinogenesis*, **31**: 927–35.
38. Nakamura, K., Ito, Y., Kawano, Y., Kurozumi, K., Kobune, M., et al. (2004) Antitumor effect of genetically engineered mesenchymal stem cells in a rat glioma model. *Gene Ther.*, **11**: 1155–64.
39. Hong, X., Miller, C., Savant-Bhonsale, S., and Kalkanis, S.N. (2009) Antitumor treatment using interleukin-12-secreting marrow stromal cells in an invasive glioma model. *Neurosurgery*, **64**: 1139–46.
40. Nakamizo, A., Marini, F., Amano, T., Khan, A., Studeny, M., et al. (2005) Human bone marrow-derived mesenchymal stem cells in the treatment of gliomas. *Cancer Res.*, **65**: 3307–18.

41. Xu, G., Jiang, X.D., Xu, Y., Zhang, J., Huang, F.H., et al. (2009) Adenoviral-mediated interleukin-18 expression in mesenchymal stem cells effectively suppresses the growth of glioma in rats. *Cell Biol. Int.*, **33**: 466–74.
42. Gunnarsson, S., Bexell, D., Svensson, A., Siesjo, P., Darabi, A., and Bengzon, J. (2010) Intratumoral IL-7 delivery by mesenchymal stromal cells potentiates IFN-γ-transduced tumor cell immunotherapy of experimental glioma. *J. Neuroimmunol.*, **218**: 140–4.
43. Kim, S.M., Lim, J.Y., Park, S.I., Jeong, C.H., Oh, J.H., et al. (2008) Gene therapy using TRAIL-secreting human umbilical cord blood-derived mesenchymal stem cells against intracranial glioma. *Cancer Res.*, **68**: 9614–23.
44. Menon, L.G., Kelly, K., Yang, H.W., Kim, S.K., Black, P.M., and Carroll, R.S. (2009) Human bone marrow-derived mesenchymal stromal cells expressing S-TRAIL as a cellular delivery vehicle for human glioma therapy. *Stem Cells*, **27**: 2320–30.
45. Sasportas, L.S., Kasmieh, R., Wakimoto, H., Hingtgen, S., van de Water, J.A., et al. (2009) Assessment of therapeutic efficacy and fate of engineered human mesenchymal stem cells for cancer therapy. *Proc. Natl. Acad. Sci. USA*, **106**: 4822–7.
46. Miletic, H., Fischer, Y., Litwak, S., Giroglou, T., Waerzeggers, Y., et al. (2007) Bystander killing of malignant glioma by bone marrow-derived tumorinfiltrating progenitor cells expressing a suicide gene. *Mol. Ther.*, **15**: 1373–81.
47. Kinoshita, Y., Kamitani, H., Mamun, M.H., Wasita, B., Kazuki, Y., et al. (2010) A gene delivery system with a human artificial chromosome vector based on migration of mesenchymal stem cells towards human glioblastoma HTB14 cells. *Neurol. Res.*, **32**: 429–37.

48. Uchibori, R., Okada, T., Ito, T., Urabe, M., Mizukami, H., et al. (2009) Retroviral vector-producing mesenchymal stem cells for targeted suicide cancer gene therapy. *J. Gene Med.*, **11**: 373–81.
49. Gu, C., Li, S., Tokuyama, T., Yokota, N., and Namba, H. (2010) Therapeutic effect of genetically engineered mesenchymal stem cells in rat experimental leptomeningeal glioma model. *Cancer Lett.*, **291**: 256–62.
50. Sonabend, A.M., Ulasov, I.V., Tyler, M.A., Rivera, A.A., Mathis, J.M., and Lesniak, M.S. (2008) Mesenchymal stem cells effectively deliver an oncolytic adenovirus to intracranial glioma. *Stem Cells*, **26**: 831–41.
51. Josiah, D.T., Zhu, D., Dreher, F., Olson, J., McFadden, G., and Caldas, H. (2010) Adipose-derived stem cells as therapeutic delivery vehicles of an oncolytic virus for glioblastoma. *Mol. Ther.*, **18**: 377–85.
52. Glover, D.J., Lipps, H.J., and Jans, D.A. (2005) Towards safe, non-viral therapeutic gene expression in humans. *Nat. Rev. Genet.*, **6**: 299–310.
53. Pack, D.W., Hoffman, A.S., Pun, S., and Stayton, P.S. (2005) Design and development of polymers for gene delivery. *Nat. Rev. Drug. Discov.*, **4**: 581–93.
54. Yukawa, H., Mizufune, S., Mamori, C., Kagami, Y., Oishi, K., et al. (2009) Quantum dots for labeling adipose tissue-derived stem cells. *Cell Transplant.*, **18**: 591–9.
55. Yukawa, H., Kagami, Y., Watanabe, M., Oishi, K., Miyamoto, Y., et al. (2010) Quantum dots labeling using octa-arginine peptides for imaging of adipose tissue-derived stem cells. *Biomaterials*, **31**: 4094–103.
56. Wang, H.C., Brown, J., Alayon, H., and Stuck, B.E. (2010) Transplantation of quantum dot-labeled bone marrow-derived stem cells into the vitreous of mice with

laser-induced retinal injury: survival, integration and differentiation. *Vis. Res.*, **50**: 665–73.

57. Rosen, A.B., Kelly, D.J., Schuldt, A.J., Lu, J., Potapova, I.A., et al. (2007) Finding fluorescent needles in the cardiac haystack: tracking human mesenchymal stem cells labeled with quantum dots for quantitative in vivo threedimensional fluorescence analysis. *Stem Cells*, **25**: 2128–38.
58. Huang, D.M., Hung, Y., Ko, B.S., Hsu, S.C., Chen, W.H., et al. (2005) Highly efficient cellular labeling of mesoporous nanoparticles in human mesenchymal stem cells: implication for stem cell tracking. *Faseb J.*, **19**: 2014–16.
59. Yan, R.H., Shan, H., Nie, L.B., Li, D., Wang, J., et al. (2010) Magnetic/luminescent quantum dots bifunctional nanoparticles labeling of rat bone mesenchymal stem cells. *Zhonghua Yi Xue Za Zhi*, **90**: 56–60.
60. Brekke, C., Williams, S.C., Price, J., Thorsen, F., and Modo, M. (2007) Cellular multiparametric MRI of neural stem cell therapy in a rat glioma model. *Neuroimage*, **37**: 769–82.
61. Anderson, S.A., Glod, J., Arbab, A.S., Noel, M., Ashari, P., et al. (2005) Noninvasive MR imaging of magnetically labeled stem cells to directly identify neovasculature in a glioma model. *Blood*, **105**: 420–5.
62. Wu, X., Hu, J., Zhou, L., Mao, Y., Yang, B., et al. (2008) In vivo tracking of superparamagnetic iron oxide nanoparticle-labeled mesenchymal stem cell tropism to malignant gliomas using magnetic resonance imaging. Laboratory investigation. *J. Neurosurg.*, **108**: 320–9.
63. Bulte, J.W., Brooks, R.A., Moskowitz, B.M., Bryant, Jr L.H., and Frank, J.A. (1998) T1 and T2 relaxometry of monocrystalline iron oxide nanoparticles (MION-46L):

theory and experiment. *Acad. Radiol.*, 5(Suppl. 1): S137–S46.

64. Bulte, J.W., and Kraitchman, D.L. (2004) Iron oxide MR contrast agents for molecular and cellular imaging. *NMR Biomed.*, **17**: 484–99.
65. Yang, F., Cho, S.W., Son, S.M., Bogatyrev, S.R., Singh, D., et al. (2010) Genetic engineering of human stem cells for enhanced angiogenesis using biodegradable polymeric nanoparticles. *Proc. Natl. Acad. Sci. USA*, **107**: 3317–22.
66. Kim, S.E., Jeon, O., Lee, J.B., Bae, M.S., Chun, H.J., et al. (2008) Enhancement of ectopic bone formation by bone morphogenetic protein-2 delivery using heparin-conjugated PLGA nanoparticles with transplantation of bone marrow-derived mesenchymal stem cells. *J. Biomed. Sci.*, **15**: 771–7.
67. Park, K.H., Kim, H., Moon, S., and Na, K. (2009) Bone morphogenic protein-2 (BMP-2) loaded nanoparticles mixed with human mesenchymal stem cell in fibrin hydrogel for bone tissue engineering. *J. Biosci. Bioeng.*, **108**: 530–7.
68. Yilgor, P., Tuzlakoglu, K., Reis, R.L., Hasirci, N., and Hasirci, V. (2009) Incorporation of a sequential BMP-2/BMP-7 delivery system into chitosan-based scaffolds for bone tissue engineering. *Biomaterials*, **30**: 3551–9.
69. Yilgor, P., Hasirci, N., and Hasirci, V. (2010) Sequential BMP-2/BMP-7 delivery from polyester nanocapsules. *J. Biomed. Mater. Res. A*, **93**: 528–36.
70. Wu, X., Zhang, J., Yi, D., Gu, C., Wei, X., et al. (2010) Enhancement of proliferation and differentiation of bone mesenchymal stem cells by basic fibrous growth factor controlled release nanoparticles. *Sheng Wu Yi Xue Gong Cheng Xue Za Zhi*, **27**: 80–5.

71. Jung, Y., Chung, Y.I., Kim, S.H., Tae, G., Kim, Y.H., et al. (2009) In situ chondrogenic differentiation of human adipose tissue-derived stem cells in a TGF-β1 loaded fibrin-poly(lactide-caprolactone) nanoparticulate complex. *Biomaterials*, **30**: 4657–64.
72. Park, J.S., Yang, H.N., Woo, D.G., Chung, H.M., and Park, K.H. (2009) In vitro and in vivo chondrogenesis of rabbit bone marrow-derived stromal cells in fibrin matrix mixed with growth factor loaded in nanoparticles. *Tissue Eng. Part A*, **15**: 2163–75.
73. Hosseinkhani, H., Hosseinkhani, M., Gabrielson, N.P., Pack, D.W., Khademhosseini, A., and Kobayashi, H. (2008) DNA nanoparticles encapsulated in 3D tissue-engineered scaffolds enhance osteogenic differentiation of mesenchymal stem cells. *J. Biomed. Mater. Res. A*, **85**: 47–60.
74. Yang, X., Walboomers, X.F., van den Dolder, J., Yang, F., Bian, Z., et al. (2008) Nonviral bone morphogenetic protein 2 transfection of rat dental pulp stem cells using calcium phosphate nanoparticles as carriers. *Tissue Eng. Part A*, **14**: 71–81.
75. Roger, M., Clavreul, A., Venier-Julienne, M.C., Passirani, C., Sindji, L., et al. (2010) Mesenchymal stem cells as cellular vehicles for delivery of nanoparticles to brain tumors. *Biomaterials*, **31**: 8393–401.
76. Lazennec, G., and Jorgensen, C. (2008) Concise review: adult multipotent stromal cells and cancer: risk or benefit? *Stem Cells*, **26**: 1387–94.
77. Djouad, F., Plence, P., Bony, C., Tropel, P., Apparailly, F., et al. (2003) Immunosuppressive effect of mesenchymal stem cells favors tumor growth in allogeneic animals. *Blood*, **102**: 3837–44.
78. Karnoub, A.E., Dash, A.B., Vo, A.P., Sullivan, A., Brooks, M.W., et al. (2007) Mesenchymal stem cells

within tumour stroma promote breast cancer metastasis. *Nature*, **449**: 557–63.

79. Yu, J.M., Jun, E.S., Bae, Y.C., and Jung, J.S. (2008) Mesenchymal stem cells derived from human adipose tissues favor tumor cell growth in vivo. *Stem Cells Dev.*, **17**: 463–73.
80. Lu, Y.R., Yuan, Y., Wang, X.J., Wei, L.L., Chen, Y.N., et al. (2008) The growth inhibitory effect of mesenchymal stem cells on tumor cells in vitro and in vivo. *Cancer Biol. Ther.*, **7**: 245–51.
81. Komarova, S., Kawakami, Y., Stoff-Khalili, M.A., Curiel, D.T., and Pereboeva, L. (2006) Mesenchymal progenitor cells as cellular vehicles for delivery of oncolytic adenoviruses. *Mol. Cancer Ther.*, **5**: 755–66.
82. Kucerova, L., Altanerova, V., Matuskova, M., Tyciakova, S., and Altaner, C. (2007) Adipose tissue-derived human mesenchymal stem cells mediated prodrug cancer gene therapy. *Cancer Res.*, **67**: 6304–13.
83. Bexell, D., Gunnarsson, S., Tormin, A., Darabi, A., Gisselsson, D., et al. (2009) Bone marrow multipotent mesenchymal stroma cells act as pericyte-like migratory vehicles in experimental gliomas. *Mol. Ther.*, **17**: 183–90.

Conclusions

DOI: 10.1533/9781908818744.167

Nanotechnology is the engineering of functional systems at a molecular scale obtained through the finely controlled manipulation of matter on atomic, molecular, and supramolecular scales. The earliest description referred to the so-called "molecular" nanotechnology, which involves precisely manipulating atomic molecules to fabricate macroscale products.[1] Subsequently, the National Nanotechnology Initiative formally established a given size threshold for the matter manipulated: from 1 to 100 nm.[2] Because of the "atomic" size of the matter and the possibility of exponential growth as a field, nanotechnology has generated a new industrial revolution, progressing from passive nanostructures to productive nanosystems.[3]

Current research fields in nanotechnology can be divided into five different approaches: the bottom–up, top–down, functional, biomimetic, and nanomaterials approaches. Bottom–up techniques seek to assemble smaller components into more complex ones and are aimed at designing organic and inorganic molecules with well-defined shape that are able to automatically arrange themselves into some useful conformation.[4] Similarly, a top–down approach seeks to create smaller devices by using larger ones, and it finds a field of application in fabricating microprocessors

Published by Elsevier Limited, 2014

smaller than 100 nm and which may also become cheaper, faster, and contain more memory.[5] Functional approaches seek to develop molecules with useful specific properties which could then be used as single-molecule components in nanodevices. Bionic nanotechnology applies biological methods and systems found in nature to the study and design of engineering biomolecular systems and modern technology, including use of viruses and lipid assemblies. Nanomaterial subfields such as nanoelectronics with carbon nanotubes and other fullerenes have given rise to nanoscale materials used for commercial and also medical applications. Display technology, lighting, solar cells, biological imaging, nanorobots applied in medicine,[6] and other innovative materials have been demonstrated, with some patents being granted to devices for future commercial applications.

The online database of the Project on Emerging Nanotechnologies estimates that over 800 manufacturer-identified nanotech products are publicly available, with new ones hitting the market at a pace of three or four per week.[7] Further applications allow tennis balls to last longer, golf balls to fly straighter, and even bowling balls to become more durable and have a harder surface. Trousers and socks have been infused with nanotechnology so that they will last longer and keep people cool in the summer. Bandages are being infused with silver nanoparticles (NPs) to heal cuts faster.

Hence, given the variety of potential applications, ranging from medical and biotechnological fields to industry and military, governments have invested billions of dollars in nanotechnology research. The USA 2014 Federal Budget provides more than $1.7 billion for the National Nanotechnology Initiative, with a cumulative investment since fiscal year 2001 of almost $20 billion;[8] the European

Union $1.2 billion with an expected market volume of €1 trillion in 2015,[9] and Japan $750 million.

The wide versatility of nanotechnology finds numerous applications in the medical field. The treatment of cancer in general and of brain tumors in particular remain a great challenge. Targeted therapies have recently been applied in different kinds of tumor, achieving good results in some cases, but their efficacy remains low when tested in brain tumors. Despite therapeutic advances – last but not least the new epigenetic-based therapies – the overall mortality is still high, especially in glioblastomas. The key points to achieve successful results in terms of morbidity and mortality involve both the multiple aberrant signaling pathways typical of the tumor progression and the presence of the blood–brain barrier acting as a gate that restricts the delivery of many chemotherapeutic agents. Of course, there are several other factors underlying the disappointing results in brain cancer therapeutics to be considered, including limited tumor cell drug uptake, intracellular drug metabolism, inherent tumor sensitivity to chemotherapy, and cellular mechanisms of resistance. Glioblastoma is a multifaceted disease, a therapy acting on a single molecular mechanism is not enough to eradicate it.

Nanotechnology provides a revolutionary opportunity for molecular treatment thanks to the engineering of nanomedicines specifically interacting with tumor cells and able to cross the blood–brain barrier. Being tumor-specifically targeted, nanodrugs will show more efficacy with fewer side effects because it is possible to use a lower dose of drug with a selective delivery to target tumor cells. Nanoparticles are able to detect the tumor at an early stage and act on cancer-specific markers. Promising in vitro results have been reported that remain to be validated in patients. Beyond the use of NPs as drugs, more focus should be placed on their role as tools to learn more about cancer biology. In fact, it is possible to take selective contrast enhancement molecules to visualize brain tumors and to study in vivo all of their characteristics in greater definition.

These unique properties are very attractive for pharmaceutical and clinical applications. Engineering of NPs for combined therapeutic and diagnostic applications (theranostic NPs) requires that the surface of NPs can be modified to achieve targeted delivery and improved biocompatibility. Compounds may also be encapsulated within the interior core of NPs for multiple functions creating multifunctional NP platforms. Platforms are able to target multiple tumor markers and deliver multiple agents simultaneously, acting as diagnostic molecular imaging agents and carrying different type of drug at same time.

In coming years, we may expect growing numbers of reports involving novel hybrid structures based upon nucleic acids, such as small interfering RNAs for the targeted silencing of the major genetic pathways associated with brain cancer development and progression, and single-strand DNA short aptamers for efficient, inexpensive, and non-immunogenic alternatives to antibodies. A great step forward has been made thanks to the conjugation of NPs with neural and mesenchymal stem cells. These stem cells possess a specific tropism for brain tumors, which makes this platform always more specific.

Although nano-derived applications have great potential, there are some concerns about their putative adverse effects on human health and the environment, as suggested by nanotoxicology research. The properties that make NPs so promising can have an impact on the ecosystem, depending on the size, shape, and chemical composition of the particle. Nanotechnology is still a relatively young field and little is known about the long-term effects of exposure to nanomaterials, especially in clearance organs such as the liver, spleen, and kidneys. Furthermore, the potential toxicity associated with the wide variety of nanomaterials available, ranges from completely inert to highly toxic. Public health research agencies, such as the American National Institute for Occupational Safety and Health, are actively conducting research on potential health effects stemming from exposures to NPs.[11] Some NP products may have unintended consequences. Researchers have discovered that bacteriostatic silver NPs used in socks to reduce foot odor are being released in the wash. These particles are then flushed into the waste water stream and may destroy bacteria which are critical components of natural ecosystems, farms, and waste treatment processes.[12] Considering health and environmental implications, it is noteworthy that inhalation of airborne NPs and nanofibers may lead to pulmonary diseases, such as fibrosis.[13] This has been demonstrated in rats breathing NPs: the particles settled in the brain and lungs leading to significant increases in biomarkers for inflammation and stress response[14] and also induced skin aging through oxidative stress in hairless mice.[15] It is suggested how some forms of carbon nanotubes could be as harmful as asbestos if inhaled in sufficient quantities. There is the need for further research aimed at obtaining a basic understanding of how they interact with the biological system in terms of biocompatibility and biodistribution, and biosafety.

Continuous experimentation for new applications of nanotechnologies has given rise to numerous patents regarding all the fields described above. The patents reported in this book concern the most recent applications in neuroscience and, specifically in brain tumor treatment, where the principal target is the overcoming of the blood–brain barrier to permit effective delivery of pharmacological compounds or to increase the efficacy of chemotherapeutic or irradiation therapies. Most of these patents found application only at in vitro or animal levels and they have not yet been tested in humans, because of the issues relative to their reproducibility in a humans without side effects. In fact, because of the great debate and for ethical reasons too, the majority of the NP applications in brain oncology remain reported for *in vitro* or animal models. Elucidating the safety of these novel materials will rely on the drawing of more accurate and broadly applicable tools.

References

1. Drexler, K.E. (1986) *Engines of creation: the coming era of nanotechnology*. New York: Doubleday..
2. *http://www.nano.gov/about-nni/what/funding*. [Accessed 14 January 2014].
3. *http://archive.thedailystar.net/newDesign/news-details.php?nid=230436*. [Accessed 14 January 2014].
4. Levins, C.G. and Schafmeister, C.E. (2006) The synthesis of curved and linear structures from a minimal set of monomers. *Chem. Inform.*, **37**(5).
5. *http://www.nanoandme.org/nano-products/computing-and-electronics/*. [Accessed 14 January 2014].
6. Ghalanbor, Z., Marashi, S.A. and Ranjbar, B. (2005) Nanotechnology helps medicine: nanoscale swimmers and their future applications. *Med. Hypoth.*, **65**: 198–9.
7. *http://www.nanotechproject.org/*. [Accessed 14 January 2014].
8. *http://www.nano.gov/nanotech-101/what/nano-size*. [Accessed 14 January 2014].
9. *http://www.epo.org/news-issues/issues/classification/nanotechnology.html*. [Accessed 14 January 2014].

10. Roco, M.C. (2005) Report international perspective on government nanotechnology funding in 2005. *J. Nanopart. Res.*, 7: 1–9.
11. *http://www.cdc.gov/niosh/*. [Accessed 14 January 2014].
12. Benn, T.M. and Westerhoff. P. (2008) Nanoparticle silver released into water from commercially available sock fabrics. *Environ. Sci. Technol.*, **42**(11): 4133–9.
13. Byrne, J.D. and Baugh, J.A. (2008) The significance of nanoparticles in particle-induced pulmonary fibrosis. *McGill J. Med.*, **11**: 43–50.
14. Elder, A. (2006) Tiny inhaled particles take easy route from nose to brain. Available from: *http://www.vetscite.org/publish/items/003129/index.html*. [Accessed 14 January 2014].
15. Wu, J., Liu, W., Xue, C., Zhou, S., Lan, F., et al. (2009) Toxicity and penetration of TiO_2 nanoparticles in hairless mice and porcine skin after subchronic dermal exposure. *Toxicol. Lett.*, **191**(1): 1–8.
16. *http://www.washingtonpost.com/wp-dyn/content/article/2008/05/20/AR2008052001331.html?hpid=sec-health&sid=ST2008052100104*. [Accessed 14 January 2014].

Index

CPSIA information can be obtained at www.ICGtesting.com
Printed in the USA
LVOW02*2258300615

444355LV00001B/11/P

9 781907 568596